Holocene Archaeology
Of the Southern Coast of Tanzania

Amandus Peter Kwekason

E & D Vision Publishing
Dar es Salaam

Supported by Prof. Felix Chami
African Archaeology Network
University of Dar -es- Salaam
P.O.Box 35050
E mail: fchami@udsm.ac.tz

Published by E & D Vision Publishing Limited
P.O.Box 4460
Dar es Salaam
E mail: Info@edvisionpublishing.co.tz
Web: www.edvisionpublishing.co.tz

Holocene Archaeology of the Southern Coast of Tanzania
(c) Amandus Peter Kwekason, 2010

Published PhD Thesis of 2010 - University of Dar -es- Salaam

Maps and photos by A. P. Kwekason

ISBN: 978 - 9987 - 521 - 67 - 8

CONTENTS

ACKNOWLEDGEMENT

*"…Courage is not without fear, but rather the judgment that something else
is more important than fear"* **Ambrose Redmoon**

This work is the result of three years of PhD archaeological field and laboratory work (2006-2008) that followed one year of theoretical preparations and supervisory chores (2005). The programme was fully sponsored by the Sida / SAREC, Sweden through the African Archaeology Network. The combination of many individual and institutional efforts that include the National Museum of Tanzania, University of Dar es Salaam and Tanzania Antiquities Department have brought this work to this stage today.

I would like to thank all those who participated in one way or another since its conceptual stage up to its final presentation. The list is long and so it is difficult to mention everyone by name.

In the first field season in Mtwara region (2006), I owe much to my field assistants, Valence Valerian, Edward Frank, John Katobes, and Didas Balimanya, who were also doing their field training in Archaeology at the University of Dar es Salaam. Besides them was Juma, a resident of Mnaida Village and our local participant, but also actively involved in surveys and endless excavations.

The second field season (2007), which entailed five expeditions, involved more people, more camping and a much wider geographical area. In Rushungi- Kiswere areas in Mbwemkuru basin (Lindi), I was assisted by Noel, C. Maligisu, Hamisi, and especially Chris, who also rescued us with his bicycle after been caught by heavy long rains.

My special thanks go to the Head teacher of Rushungi Primary School, who hosted us during our stay in Rushungi. During the longer expedition in Mchinga-Mnangole areas in Lindi, I owe much to Dr. Paul Msemwa and his driver Thomas Mpendakazi; the archaeology students: John Katobes, P. Mapinduzi and Robert Toss. In Kilwa Island expeditions, I was assisted by J. Katobes and E. Ichumbaki; Laiza (archaeology students); Mzee Yusuf and Mr. Bwanga (Antiquities staffs); Gido Laswai (AAN staff);

Bakari Yusuf, Salum Abdalah, Seleman Nasoro and Omary Yusuf (Kilwa Island residents). More thanks should go to staff members of Mnaida Primary School, where we camped for four months during 2006 and 2007 field seasons, for their social and moral support, to Rushungi and Mnangole Primary school teachers and the village administration for allowing us to camp there during field research.
I am grateful to the Director of Tanzania Antiquities Department, Mr. Donatius Kamamba, who waived fees on my excavation permit throughout this research programme and allowed free stay in the department Rest House in Kilwa Island.

Thanks to my employer, the National Museum of Tanzania, for providing study leave that enabled me to accomplish this work, including the Chairperson of the National Museum Board during the study period, Dr. D. Ndagala, The Director General of the National Museum of Tanzania, Dr. N. Kayombo, and the Director of the Museum and House of Culture, Dr. P. Msemwa. Thanks go to the members of staff who kindly gave me support that made this work successful.

Special thanks must go to Prof. Felix Chami who played several key roles in helping me to accomplish this work: First of all, he is the one who influenced the change of my career from the hard science - geological realm into Archaeology. My very first archaeological excavation was during his PhD second spell at Kiwangwa in 1991 together with Yunus, Tanzania Harbours Authority, and the Late Christine Kyembe, of Antiquities Department. He is also the Principal Coordinator for the African Archaeology Network, which sponsored my PhD studies. He is again the Supervisor of this work. All that I can do for him at this moment is to let him know that "setting an example is not the main means of influencing another, it is the only means."He also agreed to sponsor the publication of this work and accepted to be its substansive editor.

I am thankful to Prof. Terry Harrison for his moral support, exposure to field experiences and financial support. Through his financial support he made possible my trip to Calgary (Canada) where I presented my hypothetical research proposal at the SAFA Congress in 2006. This is an international conference for the Society of Africanist Archaeologists, where I got a chance to discuss with esteemed scholars and stakeholders in this topic.

Finally, to members of my family, who might have felt ignored when I was busy with this work, I have a few words for them that "action may not always bring happiness, but there is no happiness without action". Thanks for having been so patient.

Amandus Kwekason

DEDICATION

This work is dedicated to the Souls of my Ancestors, who gave me courage and showed the way into those Makonde woodlands, leading to the discovery of those long buried sites to be discussed in this thesis; to all those who gave me strength and good will; and to my beloved mother who passed away a few days before the Viva voce presentation of this work.

LIST OF ILLUSTRATIONS

LIST OF TABLES

LIST OF ABBREVIATIONS

1x2m, 2x3m, … - trench dimensions, area size

AAN – African Archaeological Network

AD – (years) After Christ

BC – (years) Before Christ

BP – (years) Before Present, i.e. 1950

C-14 – Radioactive carbon element ($^{14}C_6$)

Chikay – Chikayowa (site)

Cvx – convex (edge-shape)

Dec. – Decorated potsherds (amount, %)

Diagn. – Diagnostic potsherds (amount, %)

D-N – Deductive – Nomothetic

EIW – Early Ironworking culture

Excav. - excavation

Frag(s) – fragment(s)

HD/H-D - Hypothetical Deductive

HDN – Hypothetical-Deductive-Nomological

Incis. – Incision/ Incised (decorative technique)

Indeterm – indeterminable

Invas – invasive (nature of retouching)

L2, L3 - excavation levels, level 2, etc

'Lithics'- all stone artefacts

LSA – Late Stone Age culture

Marg – marginal (retouching)

Masak – Masakasa (site)

Mnang – Mnangole (site)

MSA – Middle Stone Age culture

Nakat. – Nakatumbatu (site)

Ngurun – Nguruni (site)

pcs – pieces

PIW – Pre-ironworking culture

Plio - Pliocene (epoch, geological time)

PSW – Proto-Swahili ware (tradition)

Punct. – Punctuations

PW – Plain ware tradition

R/material – raw material

Rushun – Rushungi (site)

T/U – trimmed or utilized piece(s)

TIW – Triangular Incised ware (tradition)

(n=20) – the actual/specific amount (=20)

Chapter **1**

INTRODUCTION

General Introduction

This book, initially produced as a PhD thesis, deals with the cultural processes that took place on the southern coast of Tanzania in the Holocene epoch. It is a period from about 20,000 years ago. The work focuses on the problems of peopling of the coast and their early cultures, its temporal change and spread over the littoral and its immediate hinterlands.

Extensive archaeological surveys and excavations of thirteen sites in Lindi and Mtwara regions, indicate that the southern woodlands and coastal Tanzania were widely populated long before the Holocene epoch. Two phases of human activities were identified for the last millennium BC in relation to the coastal settlements. The first is the pre-pottery stone-using phase with people widely spaced in small groups, most likely the continuation of Pleistocene Middle Stone Age communities. The second phase is the pottery-making phase with settled village life from about 5,000 years ago.

The earliest settled populations on the southern coast, has been identified through its Pre-Ironworking Wares (PIW). This tradition which is described as Mnaida tradition is attested to have spread to a wider area of the coast before the Early Ironworking (EIW) agricultural revolution of the first centuries AD.

EIW tradition of the southern coast is described for the first time. It seems to have strongly flourished during the 4th century AD, the time corresponding to a similar development in the interior of southern Africa. Its resemblance to the

Nkope-Gokomere EIW variants of Malawi-Zambia and Zimbabwe suggests an established axis of contact and high frequencies of interaction between the coast and the interior.

The study also established another settled coastal tradition described as "Proto-Swahili Ware" (PSW), which seems to have been confined to the southern coast of Tanzania and perhaps northern Mozambique. It flourished widely on the coast, especially between the 11th and 13th century AD. The "Swahili Ware" tradition of the coast seems to have been derived from the former, rather than a new type spreading from the northern coast.

The new data that includes pottery, lithic artefacts and charcoal for radiometric dating have been analysed. They show spatial and temporal organization of sites on the previously unexplored coast of southern Tanzania. The region seems to have been occupied for millennia by a local population, which had occasionally been moving around in response to the changing environment since the Stone Ages. Finally, the southern coast of Tanzania is placed in its inter-regional context of early cultural development of eastern and southern Africa, providing an insight for further archaeological investigations on this previously ignored region.

Abundant literature concerning the culture of the East African coast has been produced since the beginning of the last century, but with very little information about the southern coast of Tanzania. It had been argued by pre-historians in the early 1960s that in the southeast of East Africa there was a virtual human vacuum during the Stone Age (Posnansky 1961). Based on this, the early Ironworking industrial complex was assumed to have been introduced to a region where settled life was previously unknown and where the hunting and gathering economy was practised by larger part of the entire population (Phillipson 1976:3). Although until recently, accessibility of the region had been one of the major obstacles for research, the above arguments discouraged efforts to search for early settlements in the region. However, there has been a great need amongst scholars of the East African coast for information about the people and culture during the Holocene epoch. Southern Tanzania and northern Mozambique had been suggested as an area that could provide evidence of culture links between the coast and intersion to Lake Nyasa and beyond if systematically researched (e.g. Soper 1971:25).

This book is the result of research that was disired, therefore, for a long time. The research was conducted on an expanse of about 250km of the coast from Kilwa southwards to Mtwara area. An *exposé* of data from thirteen sites in Lindi and Mtwara regions, of which seven form a big cluster around Mikindani area, is discussed with reference to similar data from other parts of the East African coast. This work also adds to the recent data from the central coast and islands of Tanzania and other parts of the East African coast, especially those of Chami and colleagues cited in this work.

2

This work focused on the Holocene environment and sites found along the littoral and hinterland of the southern coast of Tanzania. The period is conceived as critical for the concentration of population in these areas which was made favourable by the changing climatic conditions in most parts of the world, resulting in early Neolithic life. However, the whole of southern Africa was assumed to have remained in the hunting and gathering mode until the alleged population explosion that led to the alleged Bantu movements after the beginning of the Christian era (Phillipson1976). It is also assumed that fundamental change in the economy was feasible after this immigration because of possession of Iron technology.

On the other hand, favourable localities in the sub-continent (i.e. margins of lakes and watercourses, the seacoasts, and the peripheral regions of the equatorial forests) could have sometimes supported nearly all or entire sedentary communities of hunting-collecting people during the Holocene. These people were able to live in this way due to permanent presence of one or more staple sources of food such as fish, animals, plants, and forest products (Clark 1970:22). It was within the same reasoning that this work was based knowing that like other parts of Sub Saharan African the Southern Coast of Tanzania base potential for ancient human sedentary life.

Although Kilwa was found to have evidence for ealy holoceme life in before it, the reported findings in the book are of later Holoceme period. That was what was found. I gained insight into the research problem, explained in this work during a short visit to Mtwara region in the year 2000, when I was invited by a geological team from the University of Bristol (UK) and Tanzania Petroleum (TPDC) researching the southern coast during September 2000. The intention was to evaluate archaeological potentiality of the areas being explored for petrollogy. I also used this opportunity to explore other areas in the region for archaeological sites.

During this reconnaissance surves, a variant of a PIW pottery known in the Kenya/ Tanzania Rift Valley region as Nderit (Bower et al. 1977; Robertshaw 1990) was discovered. Nderit tradition has been dated to between 7000 and 3,000 years BP. In the same period, Remigius Chami had found in the Tendaguru area, within the Mbwemkuru valley, another variant of a PIW culture (Chami & Chami 2001) related to the Narosura of the mentioned Rift Valley. This Tradition has also been dated to between 3000 and 2300 BP (Bower & Nelson 1978; Odner 1972).

A careful examination of the pottery from the two areas by Prof. Chami established presence of pre-Ironworking pottery in the region of southern coast of Tanzania (Chami & Chami 2001; Chami & Kwekason 2003). The above interpretations led this author to look for more Holocene sites on the Southern coast that could offer a better understanding of the cultural dynamics of the region.

The major concern of this work is about the beginning of sedentary life on the Southern coast of Tanzania and its subsequent cultural dynamics. Already a preliminary report of the region has shown the presence of early cultural remains in the form of pottery that could be related to the early farming phase of the Rift and Nile valleys. Comprehensive study of it was needed.

The work is divided into seven chapters. The first two chapters introduce the main ideas, contexts of this work, research problems and theoretical background. Chapter Three and Four are about the literature and environmental settings of this work. A body of literature about the subject, region and its geography is reviewed. Chapter Five present the fieldwork survey and excavations. Finds and discoveries of about twenty-four weeks of detailed field investigations are brought together in this chapter. Chapter Six present the detailed analyses of the finds reported in the previous chapter and their radiocarbon dating results. The last chapter is devoted to discussions and conclusion, whereby the new data are carefully evaluated against the existing body of knowledge.

The Geographical Focus

This work focused on the southern coast of Tanzania, which is also the southern coast of East Africa. The coast is part of a larger Swahili coast (see Chami 2002a), the land along the western seaboard of the Indian Ocean extending from southern Somalia down to Mozambique. Southern Mozambique and offshore islands including Comoros and Madagascar are sometimes considered parts of the Swahili coast (e.g. White 1983). Towards the interior, it is widest between the Pangani and Rufiji deltas in central Tanzania (Chami 1994).

The research area entails a relatively narrow coastal strip to the south of Rufiji River (8° S) that stretches southwards down to the country's southern border with Mozambique (10° 30' S). The width of the coast ranges from 50km up to 200km where it penetrates further inland along a broad river valley like that of Ruvuma. It is also wider on the central coast of Tanzania where it reaches the Uluguru and Nguru mountains/hills. This research managed to cover most of the important bays found on the Southern coast and parts of the hinterland up to 25 km inland. It is a woodland area, part of the southern tropical woodlands on the south of the equatorial forest. It also used to be a tsetse fly-infested zone (Clark 1980) but especially in Mtwara region slash-and-burn cultivation has brought considerable ecological changes. The population of both domestic and wild mammals that could provide animal protein is extremely low.

Based on the 2002 population and housing census conducted by the Tanzania National Bureau of Statistics, the population of Mtwara region was about 1.13million, with 90% of the population belonging to the Makonde group. The other groups includes

Makua, Matambwe, Matumbi (more to the north), Ngindo and Yao (more inland). Most of these social groups are also found in Lindi region although not necessarily in the same ratio. The population of Lindi was about 791,300 during the 2002 census. This population is made up of mainly farming communities who supplement their diet with mainly aquatic resources, as these regions were once important sisal and cashew nuts plantation area.

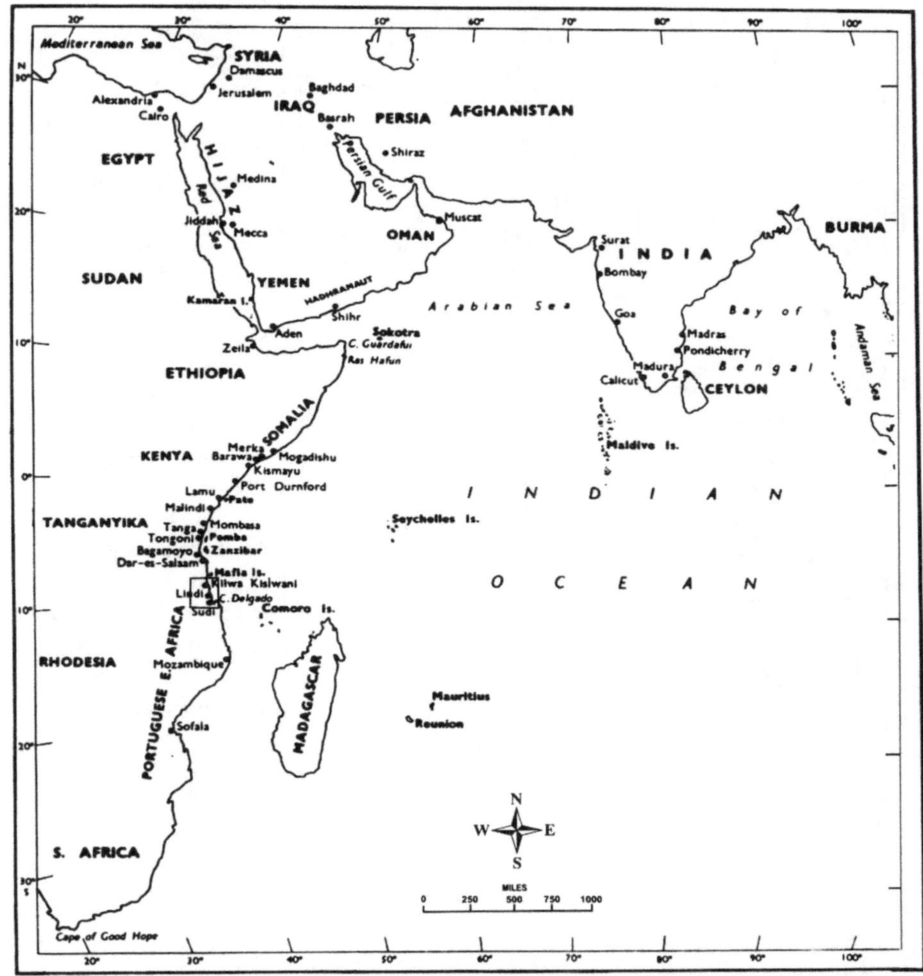

Fig. 1.1: The geographical setting of the research area (outlined), in the context of medieval towns of the western seaboard of Indian Ocean (modified from Freeman-Grenville, 1966)

Shellfish collecting has been going on along the southern coast more intensively than it has been on the central and northern coast (Msemwa 1994). Trade brought many people to the coast during the early times. Those who settled (especially from Southwest Asia) brought their religions, their customs, and their way of life to the Southern coast, some of which survived in the existing communities up to recent.

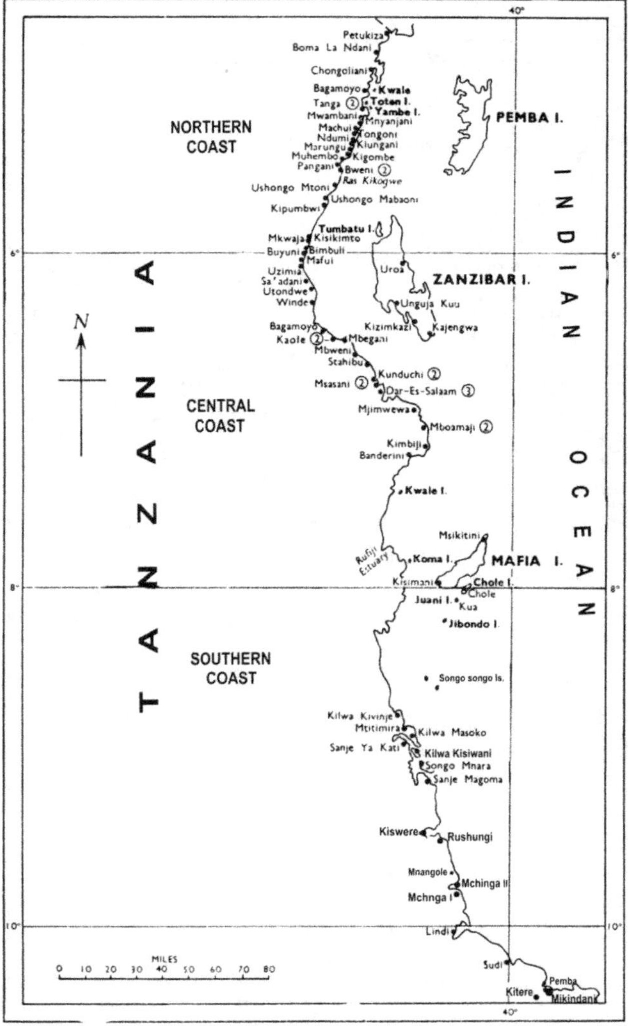

The data used in this work came from five clusters of interlinked sites possibly of the same community. Starting from the north is the Kilwa Island, Rushungi-Kiswere and Mchinga-Mnangole clusters in Lindi region; and Mikindani and Kitere in Mtwara region that stretch south to the Mozambique border. The clusters differ in size and complexity, Mikindani taking a lead in both. In Mikindani cluster, occupation sites of different periods and pottery traditions of various life spans were observed within a short range or proximal distance. It is through C-14 dating that became successful in establishing chronological sequence of the pottery traditions.

Fig.1.2: Early known famous archaeological sites along the coast of Tanzania (modified) after Freeman-Grenville 1966.

6

Problem Statement

The main research problem concerns the peopling and cultural dynamic of the Southern coast of Tanzania and also of the whole of the Swahili Coast. There are several interrelated, but mostly conceptual contradictions between the new growing body of knowledge and the extant theories. Four specific problems were strategically advanced during the formulation of this programme in order to make the work more objective and scientific.

The first problem concerns the fact that southern coastal Tanzania is said to have been unsettled prior to the EIW, meaning that farming communities started with post EIW period. It had been argued in the early schoolarship (see Chami 1994) that the coastal region, including the southern coast, which is now dominated by Bantu speakers, had no human settlements during the Stone Age (Posnansky 1961: 90). Based on the assumption above, no research has been directed at this region to look for settlements of earlier periods. As part of the southern tropical woodlands to the south of equator, tsetse fly infestation was thought to have been one of the major barriers for the spread of human population to this area (Clark 1980; Smith 1999). The beginning of human activities to the southern parts of eastern and southern Africa was then thought to have occurred during the spread of cattle herding and farming later in time.

This work aims at presenting a thorough archaeological research examination of the southern coast to prove the presence of stone-using /working communities and even settlements prior to the advent of Iron technology in the region. In Chami and Kwekason (2003), sites on the Southern coast that were related to the PIW traditions of Kenya/Tanzania Rift Valley were reported. One of the sites is Kitere in the hinterland, about 25 km from the shore. The pottery of this site shows some resemblance to the Nderit tradition of the Rift valley that has been dated to between 7000 – 3000 BP (Bower et al. 1977; Robertshaw, 1990). In the Mbwemkuru valley towards Tendaguru dinosaurs site, Stone Age assemblages have been reported (Smolla 1962). A PIW variant that resembles Narosura of the Rift valley was recently reported (Chami & Chami 2001; Chami & Kwekason 2003). This pottery tradition is dated to between 3000 – 2300 BP (Odner 1972; Bower & Nelson 1978). A short visit to the area also came up with nicely trimmed points and scrapers of Sangoan affinities. This corresponds to the lithic assemblage reported from Kilwa, assumed to correlate with the unbroken woodland belt in southern Tanzania (Isaac, G; in Chittick 1974). On the other hand, the University of Dar es Salaam field school and an unpublished report by Karoma (1982) suggested more extensive Stone culture around Kilwa. The aim of this work was, therefore, to re-evaluate the sites more thoroughly to establish the chronological sequence of the region and so demonstrate the spatial distribution of the cultures.

The second problem concerns the concept of migration in relation to the introduction of EIW into the alleged hunting and gathering Stone Age sites of eastern and southern Africa. It has been repeatedly argued, especially in Phillipson (1975; 1976; 1977; 1993; 2005), that the EIW tradition, as associated with the earliest evidences for settled villages, cultivation and ironworking technology, was introduced into the eastern and southern Africa in areas previously dominated by hunter-gatherers. It was also argued that the inhabitants of the region prior to Ironworking, they were domesticators, would have been Khoisan-speakers or hunter-gatherers (Ehret 1998; 2002:51). Also it was thought that the population was scant and without any sort of settled social organization (sees Ehret 1998). In general, it was assumed that the region consisted of small and widely spaced population, with the entire population engaged in the hunting and gathering economy (Phillipson 1976:3). The hunting and gathering tradition is alleged to have continued and dominated the region south of the Serengeti plains until the alleged arrival of ironworking Bantu Speakers in the BC/AD changeover (Phillipson 2005:212). In this connection, no EIW tradition had been discovered in the southern half of Tanzania until 2003 at Kilwa, partly due to the bias in research priority stemming from the above assumptions.

Phillipson (1976) and others have clearly demonstrated how the alleged Bantu immigrants occurred in eastern Africa introducing with it settled village life, agriculture and pottery. Before this immigration the forested and woodland areas of the coast, central and southern East Africa were considered unfavourable for a settled farming population (Robertshaw 1990). Some scholars perceived this as a "cultural lag", the result of geographic and economic factors (Sutton 1974:545). Others claimed that the tsetse fly barrier was a virtually impenetrable obstacle to the spread of a farming economy (Smith 1999). Bantu speakers are alleged to have cleared the wood using iron tools.

This work aimed to examine the cultural sequence of the Southern coast and the relationships between different cultures in the sequence. The EIW pottery has been previously argued to be the earliest pottery to be found in the region south of central Tanzania and is dated to not older than AD 300 (Phillipson, 1977; 2005). These ceramics are alleged to have spread southwards with Iron technology and Bantu agriculturists (Phillipson 1975-77; Huffman & Herbert 1994-95). This work aims at proving the presence of older pottery on the Southern coast, hence establishing a cultural relationship between the PIW variants of the Southern coast (Chami & Chami 2001; Chami & Kwekason 2003) and the EIW tradition.

The third problem concerns the argument that 'tillers of the soils', who were already inhabitants of the coast and so the founders of Rhapta and other coastal settlements that appeared in the Roman period report of *Periplus of the Erthyrean Sea* (Casson 1989), can not be related to the recent bantu speakers of the Southern coast (Huntingford, 1963; 1980; Casson, 1989; Horton, 1990; Sutton, 1990). The introduction of

Southeast Asian food crops in the sub-Saharan region through the south-eastern coast (Simmonds 1959; Posnansky 1961) was also denied in favour of Murdock's (1959) and Burkill's (1953) Sabaean lane via the Ethiopian lowlands (Fig.1.3). The time relating to *these events was considered* "too early" for south-eastern Africa and hence the southern coast, to have been settled (Chittick 1965; 1974; Huntingford 1980; Sutton 1994-95; Horton 1996). The linguistic evidences that proposed earlier settlements along the southern woodlands that extend from the southern Atlantic coast to the southern coast on the Indian Ocean (Guthrie 1962; Oliver 1966) were left unsupported by archaeological evidence. Lack of archaeological research that could provide evidence for early settlements on this coast favoured the extant Cushitic myth, that it is only in the area of modern Cushitic speakers where one finds evidence for PIW cultivators (Ehret 1982; Horton 1984; 1990; Abungu 1989; 1994-95; Sutton 1990). In areas with non-Cushitic speakers like the southern coast of Tanzania PIW traditions were not expected.

Concerning this problem, the aim in this work was to reassess the southern coast in search of evidence for early domesticating settlements dating before EIW. This would prove the presence of PIW communities, continuing of what has already been reported (Chami & Kwekason 2003; Kwekason 2007) and possibly for "tillers of soils" on the southern coast similar to what Periplus reported (Casson 1989).

The fourth problem is again conceptual; that Bantu speakers were thought to have introduced a food-producing economy, involving settled farming and animal herding (cattle, sheep, goats) in the region south of the north-eastern Tanzania (Serengeti plains) in 2nd quarter of the 1st millennium AD. The Bantu speakers were alleged to be newcomers to either virgin land or hunter-gatherers' territory or even as some scholars maintain, to Cushitic settlements, and supposedly spread the knowledge of farming, iron smelting and pottery making to the rest of east and southern Africa (Ehret 1998, Huffman &Herbert 1994-95; Soper 1982; Phillipson 1993; 2005). Besides being associated with the earliest form of settled subsistence life on the coast and southerly latitudes, their knowledge of ironworking and ancestral ceramic tradition that gave rise to EIW pottery was alleged to have been adopted from the northern Cushitic regions, in the Chad region (Soper 1971, Phillipson 1976). The EIW pottery in that respect is considered as the earliest type of pottery to the south of Serengeti / Kilimanjaro belt. For that matter, the earliest settlements on the Southern coast, although not yet demonstrated, were thought to be of Kwale EIW tradition (Soper 1971; Cruz-e Silva 1977; Sinclair 1987; Phillipson 1993).

As a corollary to the fourth problem, dates discovered recently by coastal archaeologists demonstrating that EIW sites have existed since the 1st c AD (Chami 1994) have been ignored. Horton (1990) argued that early farming communities had not settled on the coast in early times. Sutton (1994-95 pages) also argued "the date is probably just too early for the appearance of the makers of Kwale wares in the immediate hinterland, the first Bantu hereabouts.

Through this work I want to demonstrate how Stone Age people occupied the region and how PIW settlements preceded the introduction of EIW traditions. This work also aims to show how the EIW tradition of the southern coast resembles that of the woodland interior more than that of the neighbouring central coast, suggesting an influence different from the southward flow of Kwale tradition as argued by Phillipson (ibid.) and others.

Because of these problems, the need for the research on the southern coast can be seen. It is the need to demonstrate that populations adopting domestication had settled along the whole of the Swahili coast from at least 3000 BP (Chami & Kwekason 2003). It is also an attempt to show that the southern coast had been occupied since early or before the Holocene.

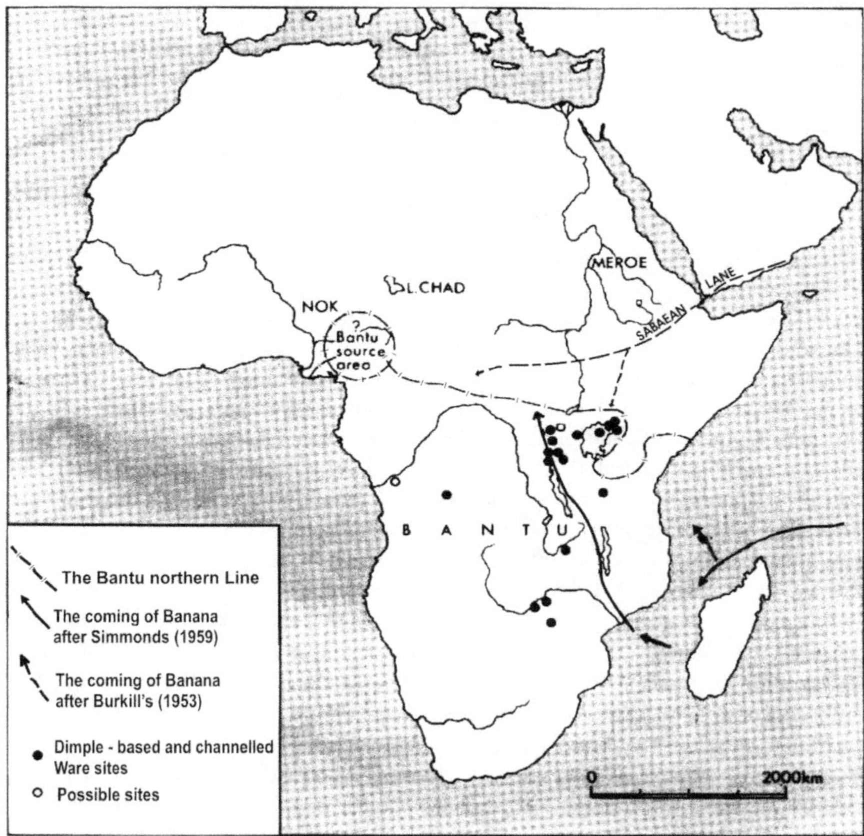

Fig.1.3: A map showing southern coast of East Africa as the entrance point for the Southeast Asian food crops before spreading into interior, after Posnansky, 1961.

Conceptual Framework

Holocene: This refers to the recent geological period that extends back to about 10,000 years BP. This is the epoch after the major Ice Age, famously known as the Pleistocene (cf. Stein & Rowe, 2006; Campbell & Loy, 2000). This epoch is considered a potential one to research for various reasons. First of all, major climatic changes that took place towards the end of the Pleistocene and more critically during this Holocene epoch, have been considered as largely responsible for the concentration of populations in many areas of the region. One of the consequences was the development of cultural periods (Clark, 1962a, b). Sutton's (1974) discussed the main phase of the wetter Holocene period, between 9th and 6th millennia BC, a period that favoured the development of aqualithic cultures in larger area of what he called Middle Africa. On the other hand, it corresponds to the time when settled farming and animal herding began in the Nile valley and the Middle East before its adoption in other parts of the world (Barich & Gatto, 1997; Cremaschi & di Lernia, 1999). Most of the African rock art is also considered to coincide with this phase (Wilcox, 1984), suggesting an increased tendency towards prolonged camping in the rock shelters. Later Holocene epoch gave way to Pre-Ironworking culture and then to EIW period.

Pre-Ironworking culture **(PIW):** This concept refers to the socio-cultural phase with clear evidence of early pottery making and sedentary village life, perhaps with some form of cultivation, hunting and fishing. It is a collective term for the Stone-working settled traditions with pottery-making skills as opposed to the Late Stone Age hunter-gatherers. The term replaces "Neolithic" in eastern and southern Africa where imprecise usage of the term or merely attribution to an ill-defined period of time is said to have hampered the study of early farming development (Phillipson 2005:172). Some other scholars have also raised questions about the use of the term 'Neolithic' due to the partial representation of the conventional features associated with the Neolithic culture in most sub-Saharan regions (Posnansky 1962; Clack 1967; Sutton 1974). While the term has been used to define sedentary village life with a farming and animal-herding economy, it has remained difficult to find real evidence for agriculture in this region. Most scholars have relied on concentrations of pottery remains and sometimes bones of domestic animals as an indication of a settled way of life, probably with some form of stable economy supported by cultivation. The skills of manufacturing and the usefulness of such fragile pottery objects have been attributed to a reasonably settled way of life (Sutton 1974:530). The PIW cultural complex is here used as a broader term representing a prehistoric culture with settled village life, pottery technology, and possibly some form of agriculture, hunting and fishing before the spread of ironworking technology. The people with this culture could have acquired manufactured iron objects through trade long before the smelting technology itself, but there is no direct evidence for that yet to provide proof, the same being true of agriculture. In this work "Mnaida tradition" is described as one variant of the PIW culture.

Early Ironworking culture **(EIW):** The concept of EIW has been adopted in this work to represent the cultural period with the earliest evidence of iron smelting in Eastern and southern Africa for reasons explained in Chami (1994, 1998). The presence of iron slag and tuyères has been used as an indication of iron working. The tradition is associated with a distinctive type of pottery, especially with thickened rims that bear bevels or flutes, sometimes up to the shoulder, found in sites scattered in a wide areas of eastern and southern Africa. The tradition first appeared in the Great Lakes region around the middle of the last millennium BC (Schmidt 1978) and by the 3rd century AD it had spread to South Africa (Phillipson 1976). Early Bantu-speaking farmers are thought to have been responsible for the spread of the EIW tradition in eastern and southern Africa (Phillipson, 1985; Soper 1971; Sinclair 1987). The term has been conventionally used as a cultural term, to replace Early Iron Age because the latter at some points conveys a temporal meaning different from that of eastern and southern Africa.

Proto-Swahili Ware **(PSW)**: This is a cultural phase similar to the period II of Chittick (1966; 1974) on Kilwa Island, of the 12th to 13th century AD. It is a distinctive culture that appears on the southern coast- south of the Rufiji River. Chittick (ibid.) had considered this tradition as of the pre-Islamic period. Although it succeeded the Plain ware tradition it was a well-established tradition along the Southern coast before the known Swahili ware tradition identified by pottery with vestigial rims, carinated form and neck or shoulder punctuation dating from 13th to 15th century (see Chami 1998). The tradition is still unknown on the central and northern coastal sites of East Africa. This tradition seems to have succeeded the Plain ware tradition in this southern coast of Tanzania that dates earlier than the known dates from the central coast of Tanzania and Mafia Island.

Culture: This refers to a consistent association of materials in a larger region thought to be contemporaneous. Childe (1958) classified cultures by their surviving culture traits – pots, implements, house forms, ornaments –known to be characteristics because they were constantly found together. "Such a complex of regularly associated traits we shall term… a culture. We may assume that such a complex is the material expression of what today would be called 'people' (Childe, 1925:15). It has been argued that archaeological culture is not the mirror image of the original (Fagan 2005). The normative parts and the superstructure of the culture are normally missing in the preserved archaeological record. It is only very few indicators recovered by archaeologists that suggest the nature of the beliefs and norms of the ancient culture.

According to Renfrew and Bahn (1996:108), archaeological assemblages and cultures are all artificial constructs designed to put order into disordered evidences.

Mnaida tradition: The usage of the term is restricted to one variant of the PIW culture; well represented by lower Mnaida hill pottery, lower Kitere hill pottery and probably lowest pottery horizon at Kilwa. Up to this stage, two radiocarbon dates are known for this tradition, one from Mnaida hill dating to the 3rd century BC and another from Kilwa Island dating to the 8th century BC. This tradition has been attributed to the Late Stone Age pottery-making settlers of East Africa due to its temporal distribution, lithic artefacts association and its occurrences below the known EIW horizon (Chami & Kwekason 2003; Chami 2006). According to S. Wandiba and C. Kusimba there is a very close affinity between this tradition and the Kenyan Rift valley LSA potteries especially the Nderit and Elmentaital types (Pers. Comm; Nov. 2009).

Chapter **2**

THEORY AND RESEARCH STRATEGIES

This chapter discusses issues related to archaeological theories and the way it has affected research in eastern Africa. The theoretical discussion will also include explanations of particular strategies adopted in this archaeological inquiry. Today's archaeology is conceived as a product of its own history, and of the intellectual climate in which it emerged (Trigger 1989a). Today, there is a superabundance of approaches that explain the importance of archaeology in bridging the past and present. The archaeological literature is awash with polemical discussions between positivists, Marxists, Structuralists, post-processualists and others, all claiming some special insight about the past (Renfrew & Bahn 1996). All these cannot be accommodated in one chapter but rather the only prominent schools of thought will be discussed.

The history of archaeology shows that archaeological thinking and approaches have been evolving over time, with new inputs being added by new scholars, with old ones changing their position in the process of adapting to social changes or needs. Sometimes people get tired or bored with the old ways of doing things; or even changes may influence the way of doing or thinking to the point of being dissatisfied with the old traditional ways. When reading through the theories, I noticed that there is a problem of "equifinality" within the theoretical body itself, similar to what Shanks and Tilley (1989b) called "meta-theoretical" problems. Some of the concepts and thoughts have lost its original meaning, such that it can now be interpreted in more than one way. Chami (2011) does not think there is equifinality in archaeology.

Previously, it was argued that patterns in archaeological records could be satisfactorily be interpreted or explained in different ways and with reference to a number of different possible processes (Hodder & Orton 1976). Like texts or words that could be interpreted differently by different people, it was learnt that a range of different simulated processes could produce similar patterns or traces in the archaeological record. In other words, most archaeological patterns, traces and processes are said to be "equifinal" in the sense that they could be expressed or formed in more that one way. What is called archaeological sites, settlement or even artefacts can be defined in several different ways and its formation and function also differ. It is becoming common for different schools to interpret the same archaeological practices differently. Differences of views have been a function of time and thinking rather than of facts. Different approaches have been suggested as time went on, in order to bring the bridging process up to date and more effective while the bridges themselves remained the same. It is clearer that a new coat of paint does not change an old house into a new one but rather it gives it a better look. Apart from the problems of equifinality, whether archaeology is a natural science or a social science, archaeology has made a tremendous contribution to an understanding of the human past through the use of scientific approaches. It has remained a discipline that mostly uses scientific methods and approaches to reconstruct objective knowledge of the past.

The Academic Essence of this Work

This work is a product of archaeology, a discipline that studies human cultural past through analyses of material remains. The author did this work using reconnaissance survey after which intensive surveys and excavations were conducted on selected areas. These surveys were executed strategically with specific research questions and expectations in mind, which are the key scientific methods for data collection. It has been argued that empirical data remains meaningless unless approached from a particular problem context (Binford 1969:271). To make it scientific, the empirical research method demands that data are collected from observations of phenomena, hypotheses formulated and tested and conclusion drawn that validates or modifies the original hypotheses (Larsen 2008). This approach was used to solve the research problems discussed in section 1.3 (Chapter One) of this work in order to objectively fill gaps in scientific knowledge about past cultures and lives of the Southern coast.

As shown earlier, the purpose was to be able to show the cultural sequences of the southern coast during the Holocene period. With careful use of the methods as stated above, in combination with data from historical documents, I become member to the group of scholars who think that through the use of scientific methods and approaches, archaeology can provide objective knowledge about past cultures, in this case – the southern coast of Tanzania.

Archaeology as Science

There has been a lot of debate about the capacity of archaeology and its methods in providing objective knowledge about the past. The debate began with what has been called 'New Archaeology' in the early 1960s. This school of thought was viewed as orthodoxy, emerging from a reaction against traditional, cultural, historical and descriptive approaches to the material past (Shanks & Hodder 1995). It does not look like a single set of beliefs or theories, but rather a diverse set of archaeologists with different approaches and beliefs, but united by a sense of dissatisfaction with the way archaeology was going and so wanting change. They wished to become more scientific and more anthropological in their doings (Johnson 1999: 15).

Actually, this dissatisfaction with the way archaeology was carried out had started earlier, in the late 1940s. Archaeologists were tired for their preoccupation with description and artefact classifications rather than providing an interpretation of past social systems and an explanation of cultural changes (Taylor 1948). There was a realization that there was no well-established body of theory to underpin methods of archaeological inquiry (Renfrew & Bahn, 1996: 441). The knowledge of artefacts had improved but without progress in an understanding of the past. In other words, the past and the present were not bridged (Clarke 1972:2). This dissatisfaction is in fact what led to a reform movement by a handful of archaeologists who adopted a logical empiricist conception of science in order to reform their discipline by making it explicitly scientific. Under the "Processual" approach of the "New archaeology" there have been many attempts to provide an underlying body of theory.

New procedures developed by the New Archaeology school, especially under Binford, were the hypothetical-deductive approaches also refered as Hypothetico-Deductive- Nomological approaches or HDN (Binford 1962; 1964; 1969). Under this epistemology, archaeology is conceived as anthropologic rather than a social-historical subject. It values explanation more than just description of sites and artefacts. It favours explanation via the incorporation of particular observations of the material past into cross-cultural generalizations pertaining to natural as well as social processes, hence the term processual. The approaches call for an explanation through explicit methodologies modelled in the hard sciences, hence "Scientism". An earlier interest in laws of human behaviour shifted to an interest in formation processes of the archaeological record, regularities that would allow inferences about processes to be made from material remains. More precisely, archaelogists of this school were becoming positivists who believed mostly, if not wholly, in scientism (Binford 1972).

The early New Archaeology involved the explicit use of theory, models and generalization. However, it was criticized for being too functionalist, too concerned with ecological aspects of adaptation purely utilitarian.

Renfrew & Bahn (1996), also Johnson (1999), have provided the reaction that followed up after the birth of the New Archaeology . The school against New Archaeology came to be termed post processual, archaeology most scholars in the school prefering non-scientific method to archaeology. Stucturalism and interpretive archaeology was advanced. These methodological debates in archeology occured mostly in Europe and America, but they affected research in Africa since most archaeologsts working in Africa were from those continents.

On the other hand during the 1970s there was a tremendous increase in the knowledge about sub-Saharan Africa, mainly due to improved archaeological techniques in surveying and excavating. It seems as if the recognition of archaeology as part of science became an incentive for the increased archaeological work. This improvement was also accompanied by the use of radiometric dating methods, which, for the first time, established chronometric dating for various ancient settlements. Improvement was also caused by many archaeologists testing their postivistic hypotheses in Africa such as that of Schmidt (1978).

Although there had been an improvement in methodologies and research strategies, the understanding of the African past remained conservative through the 1970s to 1990s. The overall scholarship was caught up in a colonial paradigm (Chami 2009). The old and conventional theories established at the beginning of the 1900s were strengthened or further entrenched, instead of being clarified or modified to match the new archaeological and historical data. Most scholars, especially linguists continued to show or view Bantu speakers as borrowers of technologies from other more civilized social groups such as Cushites or Arabs (e.g. Chittick 1965, 1974; Horton 1984, 1990; Sutton 1994-95; Abungu 1994-95; etc). Behind their resistance were racial and ethnocentric prejudices, western/foreign scholarship hegemony and limited scope of research about the past cultural processes of the region, because of the above biased notions (for conspectus see Chami 2009).

Research Strategies

The strategies adopted in this work take the views of modern archaeology (Fagan 2005:74), which combines and takes advantages of various theoretical perspectives and approaches that share aspects of cultural-historical, processual, and post-processual archaeologies. By large, it adopts an open and diverse engagement of parts, with the participation of multiple perspectives and interests (Hodder 1999).

Scientific method

The term Scientific method has been used to refer to general procedures for acquiring data, developed by scientists over many years and which has come into wide use (Fagan 2005:103). It is an approach for acquiring knowledge that is continuously self-correcting, so that the conclusions reached could be subject to repeated testing and refinement. The basic principle in applying the scientific method is the embodied notion that, knowledge of the real world is both cumulative and subject to constant rechecking (Fagan 2005). This method involves 'inductive' reasoning, which makes specific observations, and makes a generalisation from these observations. It also involves 'deductive' reasoning, which follows from hypotheses formulated through induction. The researcher postulates specific implications of the hypotheses, which are then tested with the archaeological investigations based on the specific implications of those hypotheses (Larsen 2008:16).

The sources of hypotheses are also diverse and include those from ethnographic analogies as well as from the archaeological record or data itself. The sequence of generating and testing hypotheses is said to be an unending process that can be continued indefinitely to refine and improve our interpretations. In most cases, the scientific method does not attempt to prove one hypothesis correct. In testing a series of contradicting hypotheses, one eliminates those that are incorrect and isolates the single hypothesis that are more parsimonius. It has also been argued that even Science advances by disprove, promoting the most adequate propositions of the moment while recognising that new data and better explanations will come along in the future (Ashmore & Sharer 2006: 12).

The hypotheses for this research problem came from the archaeological data (see Problem Statement in Chapter One). From the conceptual perspective, there was already dissatisfaction within Africanist School of Thought (see Chami 2006). On the other hand there was undisputed new archaeological data that suggested the presence of quite early settlements along the coast and hinterland and a significant continuity in occupation and transformation of culture (Chami & Kwekason 2003; Chami 2006; Kwekason 2007).

Under the scientific method, there are investigations in the field as well as in laboratory, to test the hypotheses, looking for the indicators that support the proposition through surveys, excavations, analyses and interpretations. Remarkably, there is almost no literature on how archaeologists come to their conclusions (see Hodder 1999).

Archaeological investigations (fieldwork, laboratory work) were organised and conducted over a wide area, from Kilwa southwards to Mtwara, covering a stretch of more than 250km of littoral and hinterland to test the hypotheses based on predefined indicators. Feasibility study was conducted all-over the stretch of the coast to establish focal points that could cater for camping and initial favourable location for a site. Surface survey was employed first to define the concentrations of sites/cluster and secondly to evaluate the sites in each cluster. It proved to be very difficult to survey every spot in most areas, especially where land is densely populated with people, or overgrown with vegetation like thickets. That is where induction plays a major role. A member of the local residents was always included in the surveying team for two reasons: they are conversant with the environment under investigation, and to assist in familiarisation and socialisation problems. As a result, the author managed to work even on restricted private lands in a convenient manner. Working along the endless network and stretches of routes, gullies and clear land, guided by local maps of the area, consumed much of the crew energy that could not match the time spent in excavations. A number of sites were defined in each area before establishing the spatial and stratigraphic relationships of the sites and then of the area /region.

Sub-surface survey was in the form of excavations. Large numbers of sites were to be excavated in each cluster. The number of trenches in each site as well as the number of clusters depended on the intensity of the search and what was required to satisfy the research objectively. Trenches of up to 16 square metres were established. The size of the trenches was mainly influenced by the observed concentrations or density of the archaeological remains in the area. Topography and the size of the working group were other factors that each time affected the choice to be made about the size of a trench. Extension of the trench to either side was unavoidable when promising evidence was revealed. The excavations also involved the searching of relevant dating materials, which in this case was for radiocarbon dating. The methods described above have been identified as main methods of archaeological science (Chami 2011).

Research design

A working plan that directed the execution of this archaeological investigation started by establishing prejudgements, prejudices and context of interpretation that furnish the conditions for real understanding of the past. This was also to ensure that the results of the work would be scientifically valid and the execution processed as efficiently and economically as possible and with minimum disturbance of the record (Binford 1964). The Ashmore and Sharer (2006:80) model of conducting archaeological research was fully utilized (Fig.2.1). The views of different specialists in other disciplines (geology, environment, chemistry, ethnology,) and others with a stake in this area of research were taken into consideration. The initial stage of the investigation identified both the problem and the geography of the region to be archaeologically investigated.

The background research formulation involved both library books and field investigation at the reconnaissance level. The potential significance of the area was evaluated in order to explore opportunities for providing new knowledge, before further input in terms of resources. The main objective was also to refine the problem being investigated and to define specific research goals with respect to the area. This stage generated hypotheses that determined the types of data to be expected that would be sought in the field. Advice from expects, whose specialist knowledge would be needed in the field and later in the analysis of the materials was also incorporated into the research at this stage. The principle supervisor and myself had a full week of reconnaissance in all of the proposed areas, from Kilwa to Tanzania's southern border with Mozambique, purposely to broadly evaluate the data universe, which otherwise does not appear without archaeological investigation. Many scenarios exist in different articles concerning the reason for this gap, including inaccessibility to the area for a longer period of the year and others logistical problems.

It was hoped that the results of these investigations would bring a considerable paradigmatic change. As a matter of procedure, the research proposal was presented to different academic bodies for evaluation. This included presentations in the department of History and Archaeology, College of Arts and Social Sciences, Directorate of Higher Learning Studies of the University of Dar es Salaam and later to the 'Society of Africanist Archaeology' (SAFA) 2006 –18th Biennial Conference held in Calgary, Alberta, Canada. The ideas and proposition of the research were discussed thoroughly and, indeed, constructive ideas and experiences were added to the research design and most doubtful areas were dispelled. Many of the archaeologists with a stake in this area of research got a chance to comment on and modify this new thinking from such an early stage, especially the participants in the SAFA conference.

During the execution of the project, progress reports were prepared as milestones, and presentations made in Kampala AAN 2006 and in Mozambique Island (Mozambique) AAN 2007 annual meetings. At all these stages, interpretations of the findings were presented and the evidence reviewed to refine the formulated programme and the understanding of the past (e.g. Kwekason 2007).

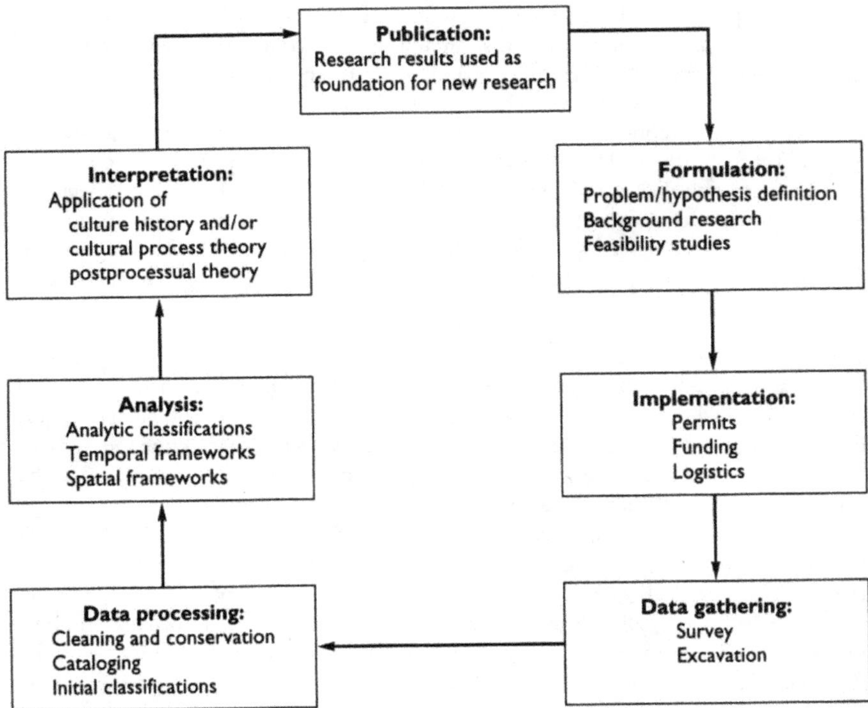

Fig. 2.1: A diagram showing the stages of research procedure (after Ashmore & Sharer 2006: 80)

Proper data collection procedures for this research were planned in advance. Locations for sampling were defined and evaluated before the execution of the research fieldwork. The data universe was established as the narrow coastal region of Lindi and Mtwara. The sample unit was expected to be the archaeological sites defined arbitrarily during the initial visit and during the reconnaissance on the region. The gathering system adopted is the "sample data gathering", situations whereby only a portion or sample of data can be collected from a given data universe (Ashmore & Sharer 2006). Under this system, a portion of every site is left untouched for future scientists to investigate, to check and refine the results obtained by this research.

The research was designed as a regional and problem oriented investigations aiming at solving specific problems through testing formulated hypotheses by using controlled and representative samples of data. A non-probabilistic sampling technique that allows the archaeologists to apply their expertise to select the sample most relevant to the research being conducted was considered suitable due to the opted extensive data universe of about 250km in length. To ensure the validity of results and make efficient use of time, money and effort, research design included a sequence of stages that guide the conduct of the investigation (Fig. 2.1).

Surveys were carried out at the reconnaissance stage followed by four phases of detailed investigations with a mobile and manageable group in form of size and skills. University archaeology students greatly contributed to the success of this exercise.

Use of analogies:

This work intends to look at the underlying cultural process, namely; the beginning and change of settlements and cultures rather than simply the material itself. It will look at pottery as one element of craft specialisation or trade, and will chart the process by which the long-term development of a market network relates to such specialisation in general. The use of analogy in archaeological interpretation of such processes is in most cases unavoidable. It is said to underpin even the most mundane interpretation (Johnson 1999:48).

Analogy refers to the assumptions made, a process of reasoning that assumes that if objects have similar attributes, then they will share other similarities as well (Fagan 2005). In archaeology, analogy involves inferring that the relationships among various traces of human activities in the archaeological record are the same as, or similar to, those of similar phenomena found among modern people. It only suggests the material culture and pattern of the modern behaviour can be used to undestand those of the past. We use modern artificial names, such as axe, to identify similar objects of the past assuming that the tools of the past were used in the same way as those of the modern time.

By assuming that the artefacts had similar uses to what we know, we also unconsciously assume that the relationship between the form and the function of the artefacts has remained static throughout the ages. By explaining the past simply by analogy with the present, we also assume that nothing new has been learned by many generations of people and that the past is not that much different from the present. Hodder (1999) pointed out that analogy is central to archaeological researches, which is much concerned with the relationship between different finds and pieces of evidence. Reasoning by analogy is important but only one step towards archaeological interpretations. It is not necessarily misleading provided that the right criterion and research strategies are used to strengthen and evaluate inferences made by the use of it (Wylie 1985).

This work made use of analogies based on the technologies, style, and function of culture as they are defined archaeologically. For instance, pointed pieces of stone are claimed to be arrowheads or tools, because plenty of similar pieces in the archaeological record have been found embedded in the bone remains of animals and people. But we have no way of knowing if similar points were part of, say, ritual activities as well as hunting. Our experiences and perceptions always control our judgments. Most analogies that are drawn from the ideas and beliefs of present-day people are probably inadequate in that manner. They do not prove or test anything but rather leave behind a number of other options or possibilities for a different interpretation.

In developing these analogies, more than one approach was used. The most common approach in use is that of direct historical analogy- simply working from the known to the unknown (Fagan 2005). Confidence in the interpretation of the past under this type of analogy diminishes as we move from historic to prehistoric times (back in time). What was done on the Southern coast is to look at the ethnographic present for the source of a theory for translating the static material found on the sites (tools, shards) into the vibrant lives of the group of people who left them there (Binford 1983: 24).

Multiple approaches

In theory, the section on different archaeological approaches were discussed in detail. It has also been learned that from the early stage of processual archaeology, the advocates were clearly bent on interpreting the archaeological record rather than just describing the artefacts and sites without relating to the social processes of that past. Controversies emerged on how to interpret the record at a later time. There was increasing diversity in the interpretation of the basic tenets. It was later admitted that an archaeological site is not really a mirror image of the material system of a society, since large proportion of the material gets some modification or disappears (Gibbon 1989).

Due to the complexity of human behaviours and so their culture, together with difficulties in dealing with the archaeological record, the most effective research seems to avoid being confined to a single theoretical approach. It has been observed (Ashmore & Sharer 2006) that no two-research situations are the same, and each tends to require a unique combination of theoretical approaches to yield the most productive results from the available archaeological data.

Together with combining elements from different approaches relevant to this research, the studies were tackled from two perspectives. The one that makes a central focus of the studies is the diachronic view. Similarly to the materialist model of culture, this view stresses development though time. The major focus is identifying and describing changes in the archaeological record over time. About twenty radiometric dates were processed from charcoal samples collected carefully during excavations to provide an absolute timeframe. The second perspective is the synchronic view, which emphasises the contemporary state of human societies with little or no time depth, and more or less adopts the ecological model of culture, the focus being on seeing the range and variety of technologies as potential clues as to how the society in question dealt with its environment as a whole. Under this, new forms or frequencies of particular material remains are viewed to have resulted from alteration of adaptive relationships between the culture and the environment.

Chapter **3**

LITERATURE REVIEW

Very little mention is made of the southern coast in most historical archaeological literature. Many scholars continue to rely on the early information about the East African coastal settlements alleged to have dated after the eighth century AD (Phillipson 2005:288). Ironically, the coast of East Africa has had a complex historical periods identified by early voyagers under different names since Pharaonic Eqypt. For instance, the *Periplus of the Erythrean Sea* and other Greco-Roman documents identified the region as Azania, and that some areas of the coast were involved in the ancient trade activities in the Indian Ocean. Lacroix (1998) suggested that some of the towns mentioned in these classical literature as belonging to the early centuries AD could have been on the Kilwa bay and Madagascar. Most of these reports have not been tested archaeologically.

In fact there is abundant literature that offer accounts of the people, settlements and other cultural aspects of the coast of East Africa, but very few of them provide a significant or direct account of these informations. However, this chapter offers a synthesis of all previous work and reports with specific account of the southern coast that provides an intellectual background to the work. It has been argued that, "Today's archaeology is the product of its own history, and of the intellectual climate in which it emerged" (Trigger, 1989). Several sources of information have been utilised, that include literature of the Classical and early Middle Ages and other academic documents, to provide the intellectual climate or context of this work.

Classical and Early Middle Age Literature

This type of literature is of ancient geographers' and travellers'. These were written for different purposes, for instance, as a guide to sailors, travellers' tales, personal documents and reports to kings. The details of this literature and its reference to other parts of the coast of Africa can be found in Mathew (1963, 1975), Freeman-Grenville (1975), Trimingham (1975) and Horton (1984). Chami (2006) has provided detail of such reports for the ancient time which clearly suggest that southern part of the Indian Ocean was known and had been settled. Chami (2009b) has also discussed the ancient and the Medieval records.

Greco-Roman documents

The earliest known accounts of the area are those given by Herodotus, Strabo and at a later time, Pliny. They seem to have roughly known the existence of the more southerly parts of Africa. Herodotus seems to have knowledge about eastern and southern Africa, 'the region that borders upon the southern Sea'. His information is of 5[th]C B.C (in Rawlinson 1964: 218). Written in the Ionic dialect of Classical Greek, his books showed that he had travelled extensively around the ancient world conducting interviews and collecting stories. He also knew of the people of the region whom he called 'long lived aethiopians' now thought to extend to East Africa (Chami 2006).

Strabo (63BC – AD 24) also recorded substantial information about the interior of East Africa, almost similar to that reported by Herodotus. He discussed geographical latitudes in relation to the position of various aspects in Africa from information compiled by previous Greek writers (see Fig. 3.1). He knew that it was possible to sail across the Atlantic Ocean around southern Africa to the Red Sea or vice versa (in Jones, 1960). He also reported of tropical conditions sourth of Equator.

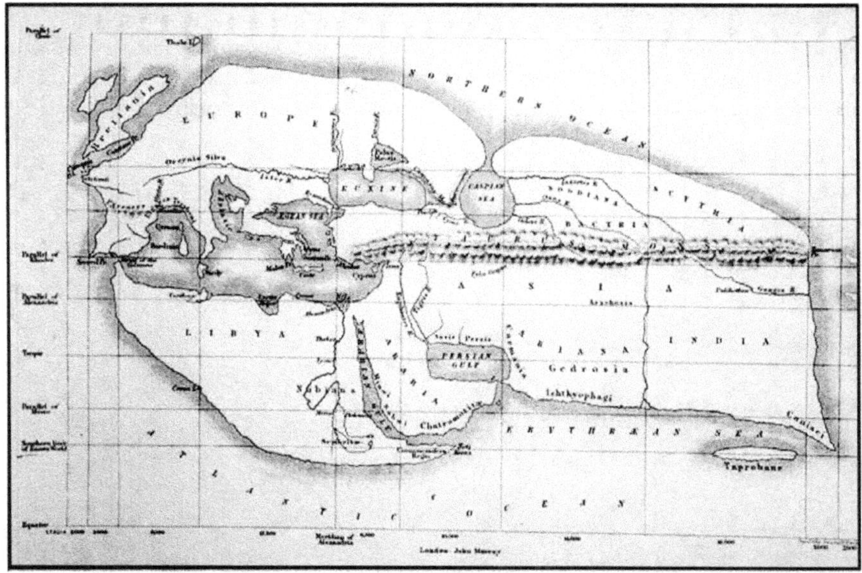

Fig.3.1: The World map according to Strabo

Pliny (BC/AD change over) in a similar manner reported about ancient sailings across the Atlantic Ocean going round southern Africa into the Indian Ocean and up to southern Arabia. Like Strabo, he pointed out East Africa and beyond in the Ocean to have been an important source of spices (Lacroix 1998: 348).

The first undisputed eyewitness account of the coast of East African, probably extending to the Rufiji Delta (see Chami 2006) is that in the *Periplus of the Erythrean Sea*, written in the middle of the first century, about AD 40-70 by unknown person. (Casson 1989). In Huntingford's version (1980), the book is given a date of AD 90-130 after Pliny's Naturalis Historial VI. The Periplus is composed of the records of pioneers intended to act as a sailing chart and travellers' handbook, a guide to the ports of Arabia, East Africa and India. (Huntingford 1980; Casson, 1989).

There are number of place references to the coast of East Africa of which the most important to this work is Rhapta. Rhapta is mentioned as the southern most known trading market. There has been a debate on the location of Rhapta. Chami (2006 :1747) suggests the region of the Rufiji Delta as the place where Phapta was located. Two main reasons he provides are that, the are is on the latitude 7^0 - 8^0 provided by *Ptolemy's Geographia* and that the region has large and many sites of the period of Azania. Actually some trade goods of the time have also been found in this region. Despite all these sets of reasons I think that Rhapta could have been located further to the south towards Kilwa and the Ruvuma river in the area of my research (see Lacroix 1998: 80). Periplus report about men of great stature, who were also agricultural (Casson 1989), who inhabited the whole coast. Each place is mentioned to have been under a separate king who had been importing from Muza /Mouza metal weapons and tools (lances, hatchets, daggers, and awls) and various kinds of glass in exchange for ivory, rhinoceros horn, tortoise shells and palm oil.

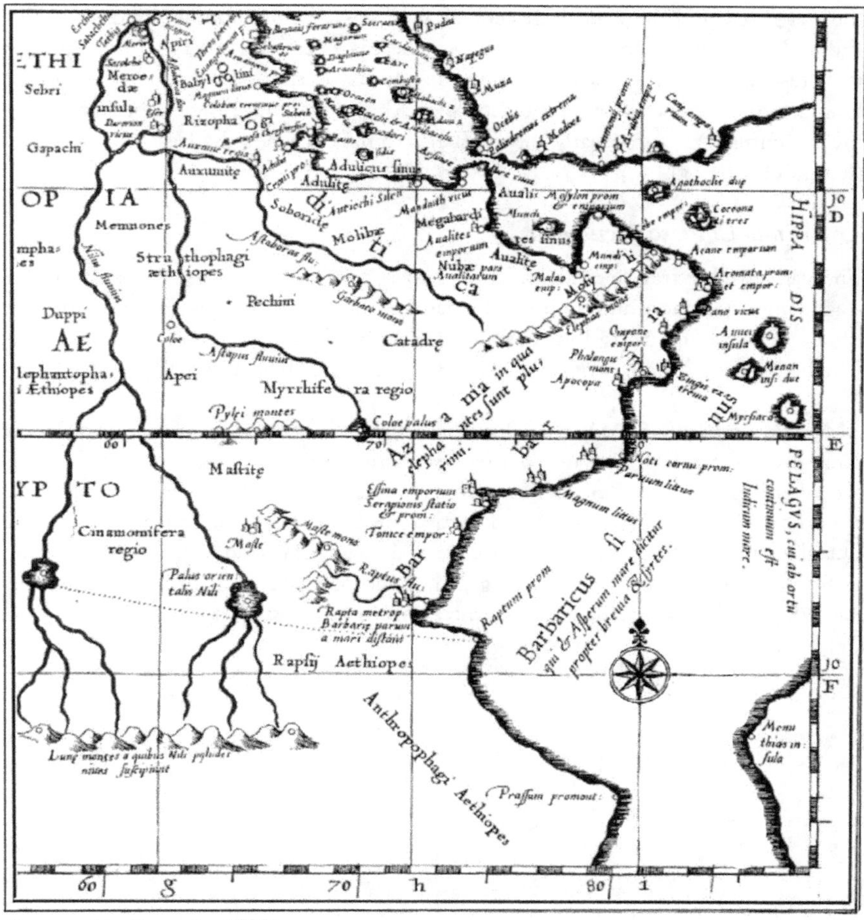

Geographia (Lacroix, 1998).

A similar account of the East African coast is that of Ptolemy's Geographia, written about a century after the *Periplus of the Erythrean Sea*. The most important information we gained from ptolemy's report are latitudes for settlements and more added trade markets. He also provides ethnical names such as identifying people at Phapta as Rufiji Ethiopes (for consepectus see Chami 2006).

From these last two documents, the impression one gets is that the southern coast of Tanzania was fully settled, with social organisations, and was involved in the international trade since or before the beginning of the first millennium AD. If one accepts Lacroix's (1998) linguistic interpretation of Ptolemy's map of Africa, the *Geographia* and the *Periplus* in general; it means that the Southern Coast of the Indian Ocean had been visted. However, whereas Lacroix (1998:114-15) suggested the presence of Bantu speakers along the southern coast at the time of Periplus, Huntingford commented that the people of Rhapta were non-Bantu speakers, since the 1st century AD is too early for the Bantu to have reached the East African coast (Huntingford 1980:147).

Arabic and Chinese documents

The early Arabic travellers identified the coast of East Africa as 'the land of the Zanj'. Important here among the documents is the account of al-Masud of the late 9th century, and of Buzurg Ibn Sharhiya of Rahormuz of the mid- 10th century in the 'Kitab al-Ajaib al-Hind', written by a Persian sailor (Freeman-Grenville 1962).

Al-Masud (c. 915-945 AD) describes the coast as inhabited by black people with their sovereign rulers. The negroid characteristics are said to have been typical of the inhabitants of the Zanj, as far south as Sofala. It should be mentioned that the region of the southern Coast of East Africa was known by Arab travellers. A trip to Quanbalu in the 9th - 10th century also led travellors to Sofala and Comoro, as far south as the Mozambican channel (see Fig. 3.3). Records available in Freeman Grenville ll (1975) of all Masud, Buzurg al Sharhiya and Idris have better been interpreted in Chami (2009b).

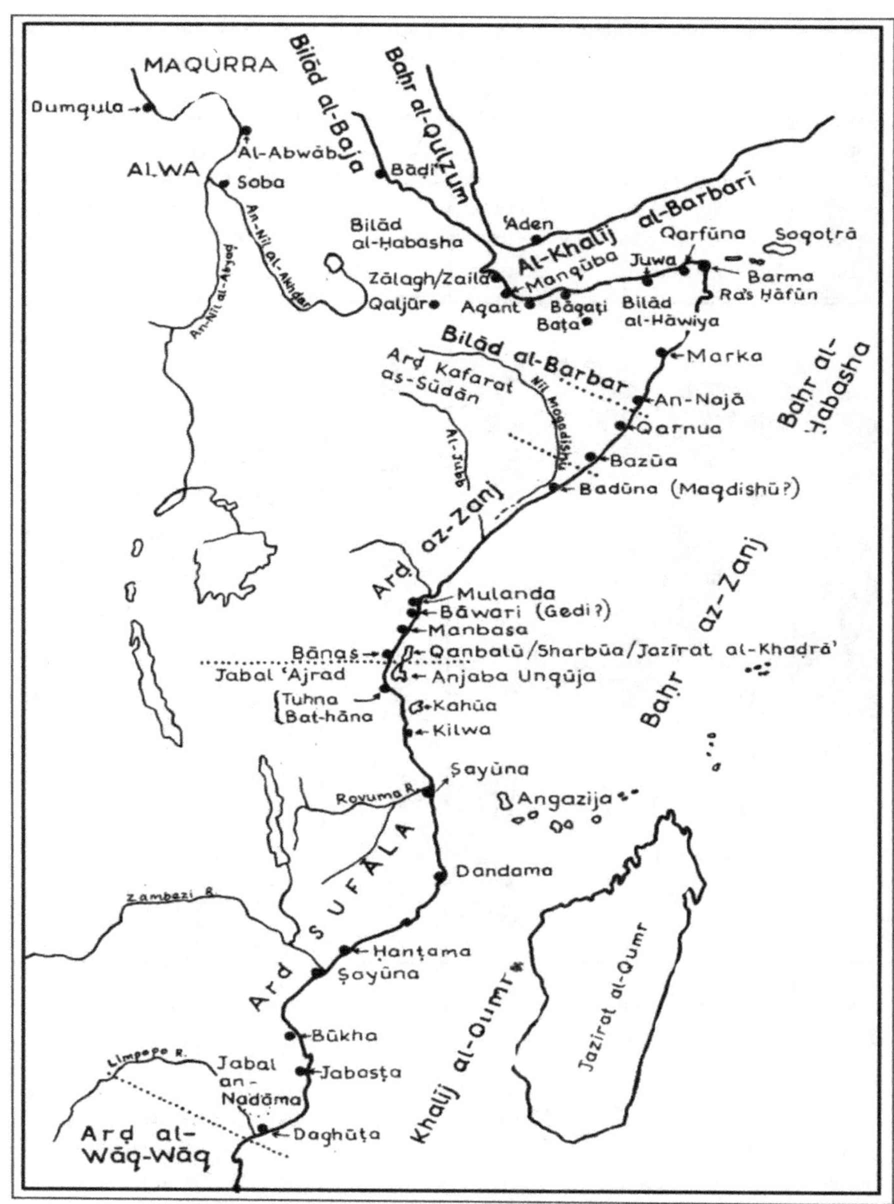

Fig. 3.3: The coast of eastern Africa according to Idrisi's map of AD 1154 (Trimingham 1975:138).

There are also a few Chinese documents, especially that of Tuan Ch'eng-Shin (AD 863) and Chao Ju-Kua (AD 1226), which, though giving accounts of the coast of East Africa, do not provide any specific information about the southern coast of Tanzania. However, despite formidable difficulties of textual interpretation, it is apparent that during the Sung times (12th c AD) and Ming times (15th c AD), there was available in China a considerable body of information relating to the East African coast as far south as Madagascar on the south-eastern coast (Fig. 3.4; Wheatley 1975: 113; 84 below). Coins of the 11th century have been excavated in Kilwa and Zanzibar (Chittick 1974 or Chami 2009b - respectively)

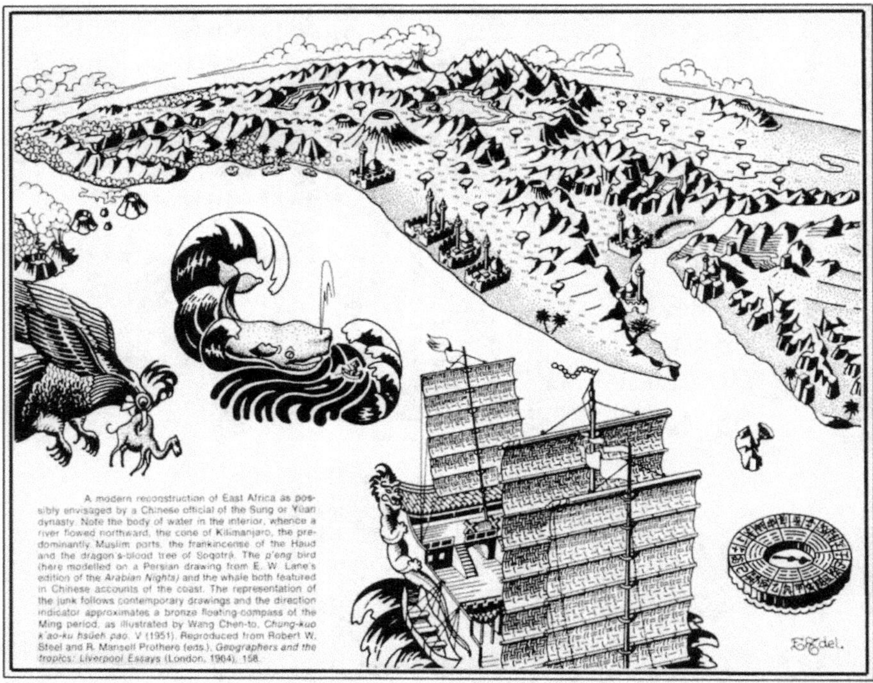

A modern reconstruction of East Africa as possibly envisaged by a Chinese official of the Sung or Yüan dynasty. Note the body of water in the interior, whence a river flowed northward, the cone of Kilimanjaro, the predominantly Muslim ports, the frankincense of the Haud and the dragon's-blood tree of Soqotra. The p'eng bird (here modelled on a Persian drawing from E. W. Lane's edition of the Arabian Nights) and the whale both featured in Chinese accounts of the coast. The representation of the junk follows contemporary drawings and the direction indicator approximates a bronze floating-compass of the Ming period, as illustrated by Wang Chen-to, Chung-kuo k'ao-ku hsüeh pao, V (1951). Reproduced from Robert W. Steel and R. Mansell Prothero (eds.), Geographers and the tropics: Liverpool Essays (London, 1964), 158.

Fig. 3.4: Reconstruction of East Africa as possibly envisaged by a Chinese official of Sung or Yuang dynasty (Wheatley, 1975:84)

Muhammad ibn Abdallah ibn Battuta (AD 1331) is the only Muslim traveller in the early 2nd millennium to give a better eyewitness description of the coastal towns of East Africa, including Kilwa (Freeman-Grenville 1962). Like Idrisi, the two offer us a better picture of the coast and its activities before the arrival of the first Europeans (Portuguese) in the late 15th century AD. By the time of Ibn Battuta's visit in 1331-1332, the population of the southern coast were Bantu speaking. The inhabitants of Kilwa, for instance, are described as of a very black complexion and with tattoo on their faces, features common among today's Makua and Makonde groups of the coast.

Prehistory and Archaeological Literature

There are quite few scholars who have conducted research or published on the southern coastal region in relation to the problems raised in this work. The region has been known from colonial times because of the Swahili period monuments of Kilwa and Mikindani and for the dinosaurs' site of Tendaguru in Lindi region.

Probably the earliest archaeological research in the region concerning early human settlement is that of Whiteley (1951) on the rock paintings of Mtwara region. Three sites were reported in Masasi district that include Masasi (near boma), Matekwa and Chitore. The paintings are of mainly geometrical designs, executed in red paint, although stylistically they resemble the late whitish designs of central Tanzania. Their location is far from the coast to the west and without any similar representation on the near coast.

Harding (1961) was the first to report isolated surface scatters of Late Stone Age artefacts that were observed around Kilwa Masoko. Smolla (1962) followed this with a study of stone tools collected from Tendaguru dinosaurs' site and other areas of Lindi. Most of these collections are still in the British Museum in London. The analysis suggested a wide spread of 'Sangoan' type of woodland Middle Stone Age tradition in the region.

At this time, excavations had already started on the Island of Kilwa. The first excavations at Kilwa consisted of a trench dug by Sir Mortimer Wheeler and Rev. Gervase Mathew. James Kirkman (1958), who accompanied the investigation at that time as in-charge of the Tanganyika Department of Antiquities, published the results of the excavations under the title of "Kilwa, Cutting behind the Defence Wall". The initial work was on a comparatively small scale and largely incidental to the task of consolidation and reconstruction of the standing ruins (Chittick 1966). It was later learnt that the wall concerned, which is situated 400m east of the Garrison/Gereza, was rather an abutment wall to support houses on the landward side (Chittick 1974).

Chittick followed with the study of Kilwa area through extensive excavations (1965, 1966; 1974). Several areas were involved in this study including Songo Mnara, Sanje ya Kati, Sanje Majoma and the Island itself. Glynn Isaac (1974) was involved to demonstrate the Stone Age remains found during this study. From Nguruni locality on the west of Husuni Kubwa, a number of stone artefacts were found lying on the eroded surface. Similar artefacts were also found in the area known as Masakasa to the east of Husuni Kubwa. The assemblage was too incomplete to be interpreted in detail, but it was certainly related to an industry, which occurs in the southern province (now Lindi and Mtwara regions) at Tendaguru (Isaac 1974: 256) in accordance with Smolla (1962).

According to Glynn Isaac (1974), Kilwa and Lindi Stone industries correlate with the belt of unbroken woodland that crosses southern Tanzania. With Levalloisian-like flakes and cores, the region was suggested to have contained a Sangoan variant of an unknown Middle Stone Age period. On the other side, Chittick (1966, 1974). established and hold-on to his argument that the 'Shirazi' population had emigrated from the Arabian states to the Southern coast of Tanzania during the 12th century AD. Although evidence for iron smelting was clear and pottery with hatched triangles and those with bevelled nicked rims were reported from the lowest level, he suggested 800 AD as the oldest age of the materials (Chittick 1966:7, 1974:317). Older dates were rejected partly because they were thought to be "too old" for the levels in which the samples were found (Chittick 1966: 9, 1974:49).

There was a major archaeological work in 1976 concerning the Stone Age of Kilwa at the area called Mgongo by the University of Dar es Salaam field school. In 1982, the field school of the University of Dar es Salaam followed up with research in Mpara (Kizimbani) hill and Nguruni areas (Karoma 1982), results which are still not yet published. Kizimbani hill was reported to have stratified succession of clearly MSA and LSA stone tools and artefacts.

On the other hand, Ehret (1998; 2002) and several others provided a linguistic explanation for the inhabitants of East Africa in general. He established that settlements and the population on the coast of East Africa reported in the Greco-Roman documents such as Periplus of the Erythean Sea and Geographia were Cushitic in nature. Bantu speakers were supposedly to have come to the coast at a much later period, probably after the 4th century AD, and are supposedly to have been a vast number of small scattering, independent agricultural communities, speaking a Bantu language and having as yet neither institutions nor the inclination to build an empire (Ehret 1998). To him, Khoisan speakers and their cultural tradition based on hunting and gathering dominated most of the eastern Africa microlithic tradition and the earliest settled tradition of the region belong to the Cushitic speakers. The latter are said to have been the dominant group over the region (Ehret 2002:170). According to Ehret (2002), the early "Chifumbaze" traditions (Phillipson 1976) now famous as EIW into eastern and southern Africa (Chami 1994; 1998) are product of grouping, the result of cultural interactions between Mashariki Bantu, central Sudanic and eastern Sahelian societies.

In the late 2000, Remigius Chami and I discovered some ceramics and stone tools and artefacts in Tendaguru (Lindi) and Kitere (Mtwara) respectively. An analysis of the pottery suggested the presence of some resemblances to the late Stone Age pottery of the Kenya and northeastern Tanzania Rift Valley (Chami & Chami 2001, Chami & Kwekason 2003). If that is fairly correct, the finds suggested the presence of early settled PIW traditions on the Southern coast of up to 3000 years old (Chami & Kwekason 2003). The communities with these traditions could be related to the previously suggested 'proto-Bantu' cultivators who had settled in the southern woodland that covers a large part of southern Tanzania up to the coast, i.e. Southern coast (Stage 2 in Fig.3.5), and who later adopted and developed Southeast Asian food crops (Oliver 1966: 366).

In 2004 and 2005, Felix Chami conducted archaeological research on the island of Kilwa aimed at discovering pre-Islamic settlements there (Chami 2005). This research involved a more comprehensive archaeological survey of the island than any previous research. Occurrences of pottery tradition that stylistically resembles the last millennium BC Narosura tradition of the Rift valley, were reported from Nguruni area (ibid.). A few other pieces suggested early contact with either India or Red Sea/ Roman Empire during the BC/AD changeover. Stone tools were also reported from this site. From Masakasa site, potsherds that were attributed to the Narosura and EIW traditions and stone tools of likely Middle Stone Age based on shape and Levalloisian flaking techniques similar to what is discussed by Isaac (1974) were encountered. All these finds suggested the occupation of the island continuously from the Stone Age period.

Emmanuel Sassi continued with excavations at Nguruni and a few other parts of the island (Sassi 2006) as part of his Masters degree. He also reported archaeological sequence that extends from Late Stone Age to the Swahili period. From this long history of archaeological research on the mentioned area, this work picked and developed a much broader research approach that established diachronic and synchronic patterns of the cultural sequences during the Holocene epoch.

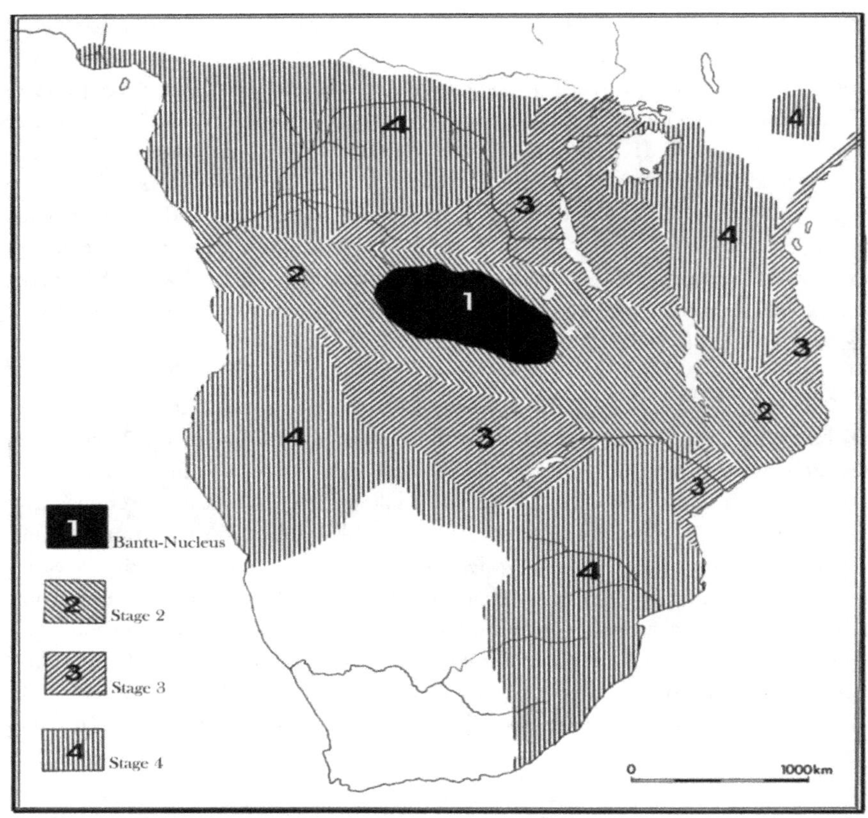

Fig. 3.5: Bantu expansion according to R. Oliver (1966:369)

Chapter **4**

ENVIRONMENTAL AND CULTURAL SETTING

Human cultures have been considered as complicated agglomerates of components such as technology, subsistence strategies and social organisation that interact with each other and in turn with the larger environmental systems of which they form a part. Human culture was ultimately seen as an adaptation to the environment (Clarke 1968; Flannery 1968; Watson et al. 1984). It has already been mentioned that the aim of this work is to deal with cultural patterns, both diachronic as well as synchronic, especially before the rise and spread of the Swahili culture (ca AD 1250). It is therefore necessary to describe environmental contexts which could have affected or conditioned the processes of adaptation and change in the area under discussion.

In archaeological theory, ecological approaches stress the study of societies within their natural environments. Cultural changes in response to environmental change became an important area of study in the mid-20th century. It was subsequently important in the emergence of the 'processual' archaeology, which primarily saw culture as an extra-somatic adaptation to the environment (Trigger 1989a, Crumley, 1994). Since then, the interplay between environment and culture continued to be a central focus in most anthropological studies.

The major influences on environmental studies came from German Friedrich Ratzel's anthropological school, with the predominating view that differences in natural habitat are often sufficient to explain cultural diversity (Ellen 1982: 2). At that time most research by the Germans was conducted in Africa. The inclination of this school of thought to diffusionism led to its downfall in the early 20th century, leaving the American cultural sphere and ecological studies to predominate (Zwernemann 1983).

The environmental conditions of the coast have been introduced here in order to provide a better understanding of the region and the ecological setting of the sites, though in the present perspective. A paleo-environmental study by itself could be conducted as a separate topic altogether to involve all dimensions and parameters of the study. It is now considered that modern conditions became established almost 4,000 or 3,000 years ago, with a decline in rainfall, and therefore, a reduction of the rain forest in most parts of East Africa (Maley 1993). This is also a period presumed to coincide with the beginning of settled village communities along the southern coast of Tanzania (Chami & Kwekason 2003; Kwekason, 2007).

Geographical Boundaries

The first chapter of this work introduced the concern with the narrow coastal strip stretching from about 8° S to 10° 30´ S. The Indian Ocean forms the eastern border of the research area, with an unstable and active shoreline. Because of its relatively shallow continental shelf, bays that offer an environment conducive for settlements since ancient time characterise the shore. Some of the bays are big and well protected from the strong convergent winds of the shore, including Kilwa, Kiswere and Mikindani bays. The archaeological record is now providing evidence of long human antiquity from these bays since Middle Stone Age.

To the southern end is the Ruvuma River with a wide low basin that extends far into the interior. The river forms the country's border with Mozambique in the south. The construction work of a bridge to link the two countries is now in process. The completion of the major construction of the Kibiti-Mtwara road has opened doors for the economic growth of the Southern coast. The region that used to be isolated from the rest of the country for about six months of the year is at last accessible throughout the year. Plateaus border the western side of the research area, whereby the Rondo Plateau is in the north of the Lukuledi River at Lindi. Makonde plateau extends south of this river. In general, the two form an enormous stretch covered with the sparsely populated Miombo woodland of southern Tanzania. Rondo plateau is less populated and more infested with tsetse flies than Makonde plateau. It seems that the latter has been affected more by slash-and-burn agriculture since ancient times, hence reducing the effect of tsetse fly infestation. The plateaus rise away from the coast up to the highest point of 900m in Rondo. Beyond these plateaus are almost open flat plains with isolated rocky hills of up to 400m high (see Fig.4.01).

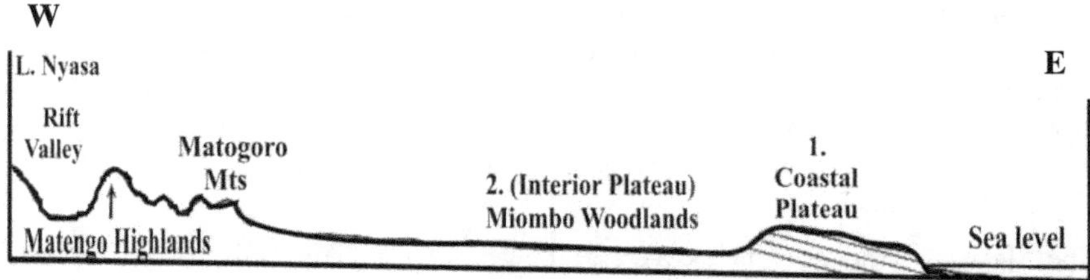

Fig.4.01: Plateaus of Southeastern Tanzania in west-east cross section (after Hickman, 1965, p.209)

The oldest settlement sites are found on the highest parts of the plateaus. There has been a change in settlement patterns from higher to lower elevation, probably a function of changing geomorphology and water levels. The youngest settlements are now closer to the shoreline. The land was once covered by the sea in the long past, resulting in the deposition of marine sediments. The coastal plateaus are the result of the uplifting of these marine beds. Recently, the sea once filled some of the river valleys near the coast. This effect is more visible at Kilwa bay, Lindi town and Mikindani bay.

Physical Features

The geography of the Southern coast of Tanzania is characterised by two major plateaus; Rondo (or Mwera) plateau that covers much of the northern part in Lindi, and, Makonde plateau that covers a larger part on the southern side up to the Ruvuma basin. The river valleys rise away from the coast up to about 900m high at the Rondo plateau (Fig.4.02). Beyond these plateaus is almost open flat plain with isolated rocky hills of up to 400m in height, remnants of formerly truncated parts of the plateaus formed by the active erosion along the coast. Perennial rivers that descend into the Indian ocean at this part of the coast after Rufiji is Matandu, to the north of Kilwa Kivinje, and Mavuji, that joins the sea at Kilwa Masoko, and the Mbwemkuru – the river with five bridges. From Mchinga to Lindi there are two more rivers, The Munimbira that joins the sea at Mchinga bay and Lukuledi on Lindi bay. Mambi River, further to the south, is formed by Lake Kitere, which is fed mostly by the Namitambo River. From the lake to the sea is 42 km. Mbuo is another river in the south with Mkundi and Matumnudi as major tributaries. Mambi and Mbuo merge 9 km before entering Sudi bay. To the east of Mikindani after Mitengo, is the Likonde River also known as Miseti downstream where it enters the Mikindani bay on its eastern wing at Miseti. The last river is the Ruvuma/Rovuma in the extreme south, forming the country's border. Among these rivers, Rufiji, the largest, is navigable for only few kilometres into the interior, followed by Ruvuma and Mbwemkuru, which are also not very useful for navigation. In the early 1917, the British colonial government made attempts to sail a ship to Utete along the Rifiji River with difficulty (Lane 1945). Utete is about 60 km up the river towards the interiour.

The northern part of the coast is thinly populated. It is covered with tsetse fly-infested Miombo woodland. The southern section, especially the Makonde plateau, is surprisingly thickly populated. The soils are poor and without water. The main crops grown in most of the area are cassava, rice and millet, and cashew nuts as a cash crop. From time to time, Rondo plateau to the north of the Makonde plateau has been exploited for timber.

The Makonde plateau, which covers a large part of the research area, comprises the country between the Lukuledi valley in the northwest and the Ruvuma valley in the southeast, bordered in the northeast by the western shore of the Indian Ocean. In the southwest, the high escarpment overlooking the Masasi peneplain borders it. The plateau is approximately 9500 km² and reaches an altitude of about 800m in its highest northwestern corner. Most of the early settlement sites are found on the higher flats and edges of the plateau. Although most of the modern settlements are now concentrated on the shore and on watered river valleys, this trend seems to be a recent phenomenon that needs to be examined.

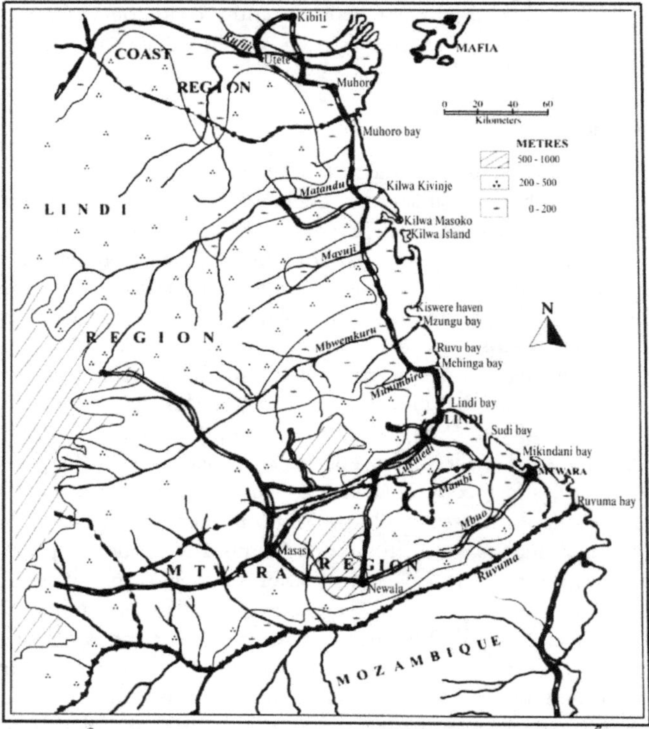

Fig. 4.02: Physical features of the southeast Tanzania (modified) from L. Berry 1977:24-25 (in Hathout, A.S.1983)

38

The eastern end of the plateau, which coincides with the coast, is much lower, varying from 45 to 90 m above sea level. The degree to which dissection has taken place differs, with the eastern or coastal end deeply cut by various valleys running into either the sea or carrying tributaries of the Ruvuma and Lukuledi rivers.

The presence or absence of water plays an important role as one of the controlling factors of the environment. On the Makonde plateau, proper water is virtually non-existent, being found only in one or two seepage lines. One of these seepages runs north south along the eastern limits of the higher plateau. The other runs west-east, and coinciding with the deep furrow of Kitangari valley (Gillman 1945). A few isolated water holes are also to be found, but their supply is precarious. The main source of water for the population is actually found below the edges of the plateau, at the foot of the scarp and in the valleys of such rivers as the Ruvuma, Mambi and Mbuo. Thus, the location of the plateau's water, entails a very large expenditure of labour in supplying the barest domestic requirements, since many villages are located far inland or at the edge of the plateau.

In trying to answer the question as to why most villages and hence most settlement sites are not found in the valley or at the foot of the scarp, several explanations were offered. One very interesting explanation is related to 'taboo'. In the region there exists a myth that the original home of the tribe's progenitor was once at "Mahuta" on the plateau. It happened from there, when he took his wife down to the Ruvuma, their first child was stillborn. The second pregnancy was when they were down on the Mbwemkuru River, but again the child was born dead. The myth continued that the third child was born after they had returned to the plateau; this time they were lucky and it survived. Henceforth, according to the mythical instruction, nature advised their offspring not to live in the valleys or near "great waters" as these were abodes of sickness and death. This, and maybe other explanations persuaded the Makonde people to live on top of the plateau, far from water sources and seashores.

Geology of the Southern Coast

The coastal plateau is formed by unconsolidated series of sedimentary rocks, mainly limestone overlain by sandstone and grit, part of the continental late Neogene (Pliocene/Pleistocene) sands. On the west, the plateau consists mainly of hard sandstone and grit overlying the old peneplain and its inselbergs carved out from the gneisses, schist and crystalline limestone of the lower basement complex. To the east, it is made up of Tertiary limestone, soft sandstone and sands resulting from the circles of denudation that began at the last uplift of the plateau during the Plio-Pleistocene.

The shore is characterised by marine terraces and a shallow continental shelf, which forms a shallow marine environments for different food resources. Fringing the coastal plain are beach ridges, mangrove swamps and deltas. Raised coral reefs form the headland east of much of the shore and peninsulas up to a height of 10m. The shallow continental shelf also made it possible for the growth of coral polyps, some of which protruded as platforms and reefs above water level in the late Plio-Pleistocene times to become smaller coralline islands. Kilwa and Songo-Songo are among the coral–formed islands on this southern coast. Kilwa Island with an area of about 64 square kilometres is a coral platform that grew over the continental shelf in an estuary environment. Terrestrial sediments with older quartz pebbles, white and red sands and clays from elevated hills of Mpara and Singino filled this estuary before its uplift during the Pleistocene. A thin layer of soil has been growing up with time (Fig.4.03).

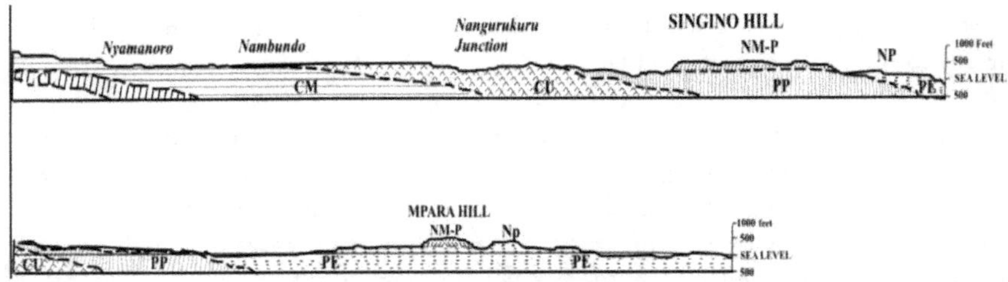

Fig. 4.03: Lithology of northern area of Southern coast: Kilwa section (source: Geol. Survey Div. Dodoma, 1963)

<u>Legend:</u>

Np   –   Grey sands with occasional small areas of red sands and low loose quartz pebbles: Outwash and reworked sands from the white and red sands of the higher Singino and Mpara hills (Pleistocene).

NM-P –   Coarse white sands with quartz pebble beds overlain by red sands or red and grey sands with quartz pebble bed, at the base of the red horizon (Pleistocene-Miocene.).

NM   –   Massive blue-grey clays and silty clays with thin detrital limestone (Miocene)

PE   –   Mudstones and marls with thin sandstone and nummulitic limestone (Middle Eocene)

PP   –   Mudstones and marls with thin calcareous sandstone and detrital limestone (Paleocene)

CU   –   Green-grey and olive grey mudstones with thin calcareous sandstone, concretionary limestone and sandy limestone (upper Cretaceous.)

<dl>
<dt>**CM**</dt>
<dd>– Blue-grey mudstones with thin sandstone and in the northern section thick basal conglomerates (Middle Cretaceous).</dd>
<dt>**CL**</dt>
<dd>– Blue-grey mudstones and Septarian concretions with minor limestone and sandstone (Lower Cretaceous)</dd>
</dl>

The unconsolidated sandstone capping the plateaus is mostly massive, brick-red in colour and of fine grains, that become more orange than red at the lower levels. Its actual thickness is unknown, but varies from place to place (Aitken 1961: 77). It contains conglomerate-like gravels on some of the lower watersheds. The sandstone has been named "Mikindani" Beds/sands, with the name derived from the section near Mikindani town on the eastern exit to Mtwara Township (Fig. 4.04). At some points these Mikindani sands get confused and misinterpreted as the "Upper Makonde Beds", which have a deceptively similar look. The latter, been older and part of a larger Makonde facies are rather unfossiliferous, reddish to reddish purple and variegated sandstone with intercalations of Marl, forming up to 30m thick. They are originally continental sediments of Upper cretaceous in time but not well dated. They appear concomitantly exposed as patches all the way from Nangurukuru to the south of Makonde plateau.

Fig.4.04: Mnaida hill with best preserved Mikindani beds. The lower section is erosion of these unconsolidated beds at Pemba site.

Climate of the Region

The earth has been in an interglacial period, which is the Holocene epoch, for about 11,000 year now. It has been suggested that the next glacial period would begin at least 50,000 years from now, even in the absence of human-made global warming (Berger & Loutre 2002). However, in the 20th century, sea levels rose by an estimated 17cm, and conservative global mean projections for sea-level rise between 1990 and 2080 range from 22 to 34cm (UN-Habitat 2008). In this UN-report, it is argued that Oceans are expanding due to the rise in sea-levels, with an increased flooding in coastal cities including the coast of eastern Africa. "Orbital forcing" is argued to be responsible for such a major change (Berger & Loutre 2002). This is the effect on climate of slow changes in the tilt of the earth's axis and shape of its orbit that affects the total amount of sunlight reaching the earth by up to 25% at mid-latitudes like our region. This effect has a major impact on the climate in general.

During the Holocene epoch, there were fewer climatic changes in the southern half of the continent than in the northern half. This is mainly due to the smaller continental masses and larger water surfaces (big oceans) of the southern hemisphere as compared to the northern hemisphere, hence a little more stable climate in the southern regions. A warmer and wetter conditions are supposed to have prevailed on larger area of the region during the humid periods from about 11,700 to 4,000 years ago (Lonnie et al. 2002). The final drop in temperature was enough to effect changes in the zoning of vegetation, but not to the point of glaciations (Willcox 1984:17). Higher mountains like the Kilimanjaro and Ruwenzori seem to have been capped with more snow than in the recent times. From about 4,000 years ago, African lakes level dropped enormously as conditions became cooler and drier (Lonnie et al. 2002: 591). The regional coincidence of low lake levels, reduction of ice cover and deposition of a thick layer of dust on Kilimanjaro summit, and societal upheaval, suggest a large climatic event centred at ~4,000 BP that lasted for about 300 years (ibid, 592). The ice field on the summit of Kilimanjaro is said to have been smaller during this greatest historically recorded drought in tropical Africa, than it is today. It now seems that sedentary life started to appear in the woodlands on the southern coast of Tanzania as this worst drought condition ameliorated, coinciding with the beginning of modern conditions in the region (Maley 1993).

The southern coast of Tanzania has two climatic seasons: a hot and humid rainy season between November and May, and a cooler, less humid dry period from June to October. Annual temperatures are fairly high, as determined by altitude and seasons. The highest monthly mean temperature ranges between 27° C and 28° C close to the coast, while the lowest is 22° C in the interior during July (Nyoni, 1998). Relative humidity ranges from 79 % – 89 % on the coast, while towards the hinterland to the west, it is commonly 10 % lower.

Regions south of the Rufiji River receive only one long rainy season (Fig.4.05). To the north of the river there are normally two rainy seasons. In general, there is less rainfall on the southern coast than in the north. The driest periods are between the months of June and October (Fig.4.06). As a matter of fact, the southern coast experiences more recurrent famine years than in its northern counterpart. Besides having a single rainy season, 69.8 % of the annual rainfall on this coast falls between November and March. This coincides with the hottest weather conditions, with a mean monthly temperature of more than 27° C (Fig.4.07). Due to this factor, much of the rainwater gets lost through evapo-transpiration, leaving behind a lesser amount of rainwater. Thus, the actual amount of rainwater available for plants or agriculture is inadequate, sometimes even for those drought-resistant crops.

The interplay between the shift of an overhead sun along the tropics and its associated low pressure belts against latitudes, altitudes and the relative distribution of land mass over water bodies subject to the rotating earth, has created westerly and easterly deflected movements of winds along the coast, causing complex rainy seasons and associated periods of drought between the northern and southern coasts of East Africa (Kwekason 2002). Two main wind systems have a strong effect on the climate of the region, which are the Northeast and Southeast Trade winds (Fig. 4.08). Where the two winds merge, including some parts of the northern coast of Tanzania, it normally receives heavy rains from conventional storms.

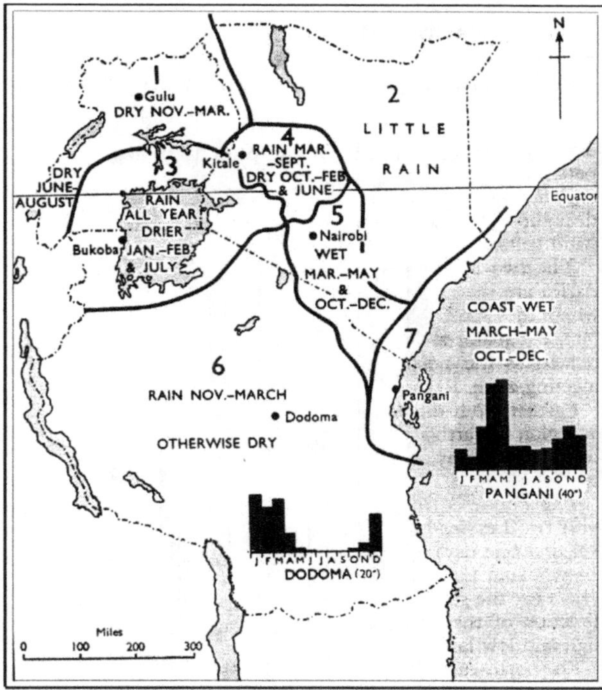

The two periods of heavier rain on the coastal region to the north of the Rufiji River (Fig.4.05) are related to the northward and southward movement of this merging convergent zone. The peaks of these heavy rainy seasons come closer and closer, the further away from the equator, towards the south until it becomes one peak on the southern coasts with only one long period of heavy rains.

Fig.4.05: East African Rainfall regions and Patterns, on northern and southern coast (after Hickman, G.M. 1965, pp. 8).

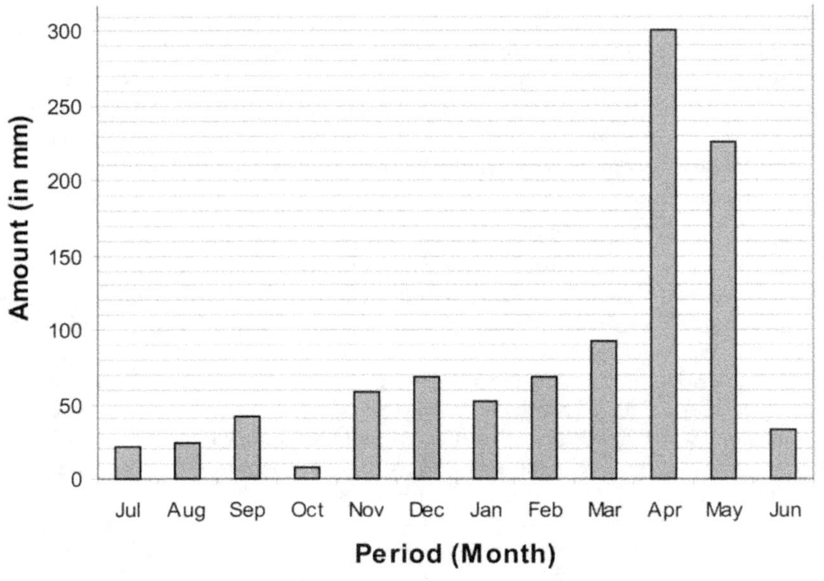

Fig.4.06: Mean monthly rainfall in Lindi region for the past 15 years. (From Lindi meteorological department)

Mean Monthly Temperature in Lindi for the past 15 years

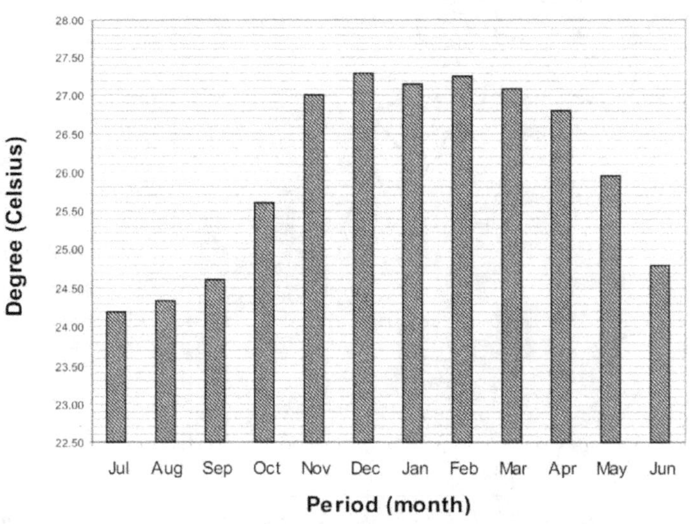

Fig.4.07: Mean monthly Temperature in Lindi region (from Lindi meteorological Department)

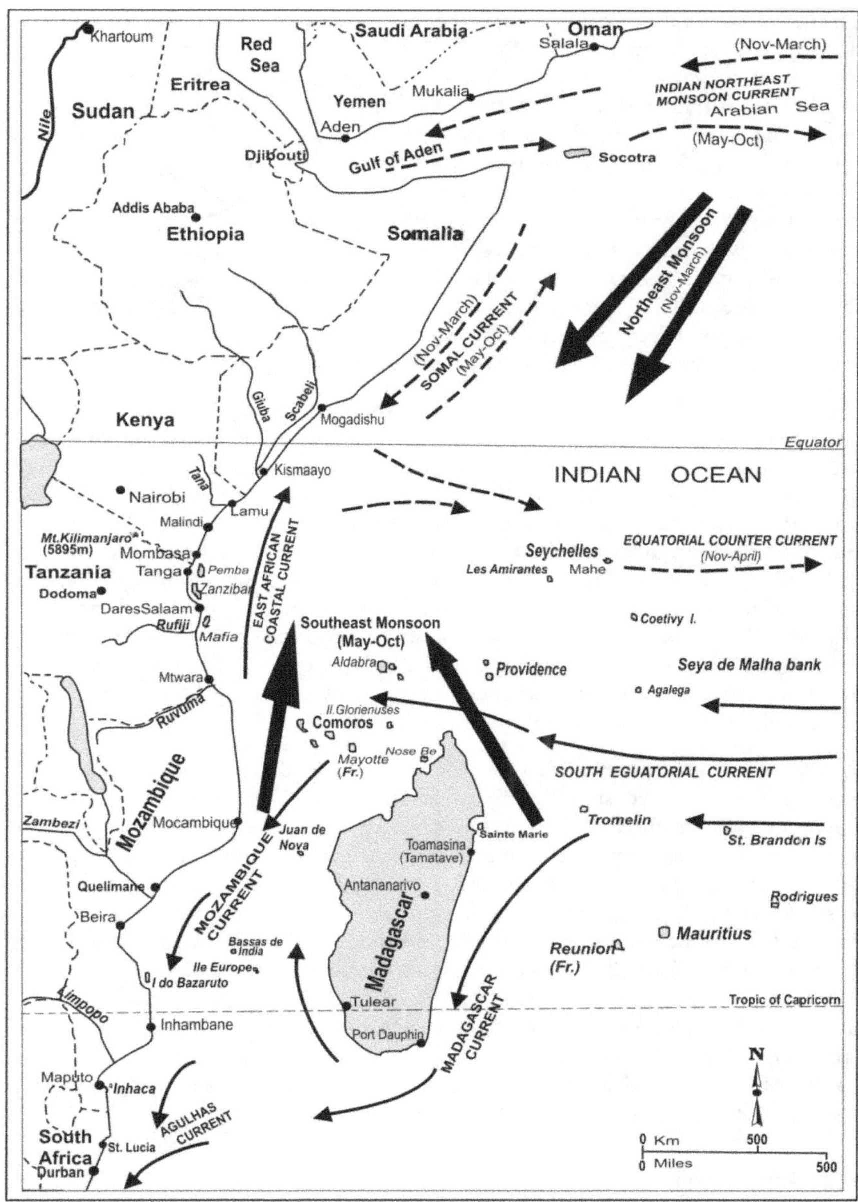

Fig.4.08: Monsoon winds and Ocean currents on the western Indian Ocean (simplified) after Ngusaru 2002: 13.

The monsoon winds have been among the most crucial climatic factors related to the development of the ancient coast of East Africa. The winds bring rains and stormy weather to the coast, a different effect from that caused by the major Northeast and Southeast trade winds. Alternating with the position of the sun, the northwest monsoon blows between November and February (Fig.4.08), which is a period of fair weather that brings precipitation to the coast south of the equator. By March, when the sun is on the equator shifting northward, the accompanied low-pressure belt brings heavy rains to the northern coast due to the double effect of the merging and converging wind systems. The southwest monsoon occurs from May to August, with cool winds, which leave no rain on the coast after blowing over a long passage over the ocean as Southeast Trade winds. The effect of the Southeast Trade winds, which are the strongest winds, prevents the second rainy season from having a significant effect on the Southern coast by driving all the rains to the north (Kwekason 2002).

Another aspect in which the winds are important is their commercial application. Since the EIW period, the monsoon winds have suppoted sailing along the Indian Ocean, bringing trade to and from East Africa (Chami 2002b). The knowledge of how the winds operated became an absolute necessity for the trade links between East Africa and the northern states of the Mediterranean, Red sea, Persian Gulf and other parts of India and Far East. The interactions and trade along the western seaboard of the Indian Ocean were made possible mostly by sailing with the help of the winds. Discussion on how different traders exploited the winds appears in Horton (1984), Casson (1989), to mention just a few. It was the monsoon winds of the Arabian Sea and western Indian Ocean that blew from the northeast during the winter, and then conveniently switched to the southwest during the summer, which made both routes (to- and from East Africa) possible.

The monsoon also determined the trade movements along the coast. The winds reached the southern coast of Tanzania in January having blown over its greatest aerial extent. From there, the winds are no longer steady. The movement further south had to rely on the Mozambique Ocean current. Unfortunately, during the return trip from the Mozambique ports, the traders had to sail against these strong currents to the western coast of Madagascar where they entered the north flowing currents which carried them to the Comoros Islands. From Comoros, the southwest monsoons are strong enough to drive dhows to the coast of Tanzania. By using these, traders from the north interested in trading as far south as Sofala, had to stay in East Africa for about a year awaiting the reversal of the winds. That in fact was the major and best opportunity for the Kilwa town to become the major trading centre, connecting the north to further gold-rich southern regions with a much shorter time (Chami 1994; Casson 1984; McClanahan 1996)

Vegetation Cover

Mangrove forests along the coast, especially from the Rufiji delta stretching south to Kilwa, some patches further south, the "Makonde thickets" on the plateau, the

Miombo woodland on the plains and some parts of the northern plateau, characterise the vegetation of the southern coast of Tanzania, which is supported by different soil types of the coast (Fig.4.09).

The plateau has, or at least had, forest cover with heavy undergrowth on the top but only patches of undergrowth on the slopes. At the extreme, there is rolling grassland with small areas of thicket, especially on the lowlands. Generally, open woodland with some thicket occurs over most of the remainder of the area, except the uncultivated stretches of alluvial soils where high grasses grow.

Swampy forests occur in the lagoons, estuaries, creeks and silted parts of the shore. Mangrove forests are typical of these places and have been intensively exploited by coastal communities since ancient times for their hard wood which is suitable for sea-moving vessels and other export trade. On the other hand, they create conducive environment for a wide range of shallow marine food resources, including varieties of shellfish, molluscs and other protein sources (Msemwa 1994).

The vegetation of the plateau in its present stage mainly consists of impenetrable secondary deciduous scrub communities, often referred to as "Makonde thickets". Studies of the floral composition and the succession relationships of the thicket communities have not yet been completed, but it is fairly clear that these thickets are a man-induced sub-climax, the result of clearing the original woodland formation followed by alternations of cultivation and regeneration by secondary scrub. Wild animals in these woodlands are very rare; probably an impact of tsetse flies infestations and/or over-hunting.

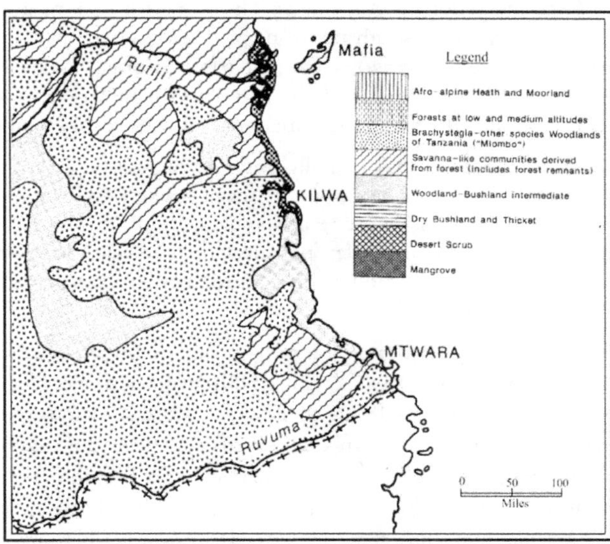

Fig. 4.09: Vegetation of Southeast Tanzania, abstracted from E.W. Russel's (1977)

People of the Southern coast have been involves in cultivation from time to time. The famous Makonde thicket-operating system of farming is well known (Gillman 1945). It involved slash-and-burn type of agriculture, whereby after about four farming seasons, the land is left to lie fallow and a new thicket field is cleared. By this method, extensive area of natural vegetation becomes affected, leading to what is called man-induced sub-climax. Ecological variations have definitely affected the settlement and economic opportunities on most of the southern coast of Tanzania. Mangroves and other forests have suffered the consequences. Climatic conditions have been blamed for affecting not only the aquatic resources but also vegetation and other economic opportunities, but in fact, it is a human induced effect.

Agricultural production on this coastal area has been so unreliable (Nyoni 1998; Msemwa 1994; Hickman 1965; Gillman 1945), that shifting in terms of settlement patterns has been one of the common practices. It is through this line of practice that a large area has become deforested. Most people have resorted to using natural resources for their survival, including intensive harvesting of the mangroves and Miombo woods. Generally, there is a higher incidence of drought than to the north of Rufiji, which is about 8° south of equator, leading to a poor regeneration of the forests and thickets.

Soils and Sustainability

The soils of the plateau are mostly light sandy loam, pale orange in colour, with stiffer textured and darker coloured subsoil, the series being very free- draining. In addition, there are small areas of almost pure sands and on some of the lowest plateau ridges near the coast are found reddish sandy loams. The Rufiji valley is mainly filled with alluvial clay and silts. Towards the south are strong brown to pale yellow unconsolidated loamy sands with some lateritic horizons. Larger parts are dominated by a dark red or red loamy sands (Fig.4.10). Large parts of the Kilwa areas, the flat areas to the south of Lindi town and lowlands further south, are enriched with coastal sands.

The fertility of all of these soils is generally not very high, as is apparent from the fairly rapid decline in crop yield within two or three seasons following the clearing of secondary vegetation. The moisture-retaining capacity of all but the white sands must however be reasonably good, as crops can withstand fairly long spells of dry weather, and the coppicing stumps left on the cleared land remain green throughout the seven-month dry season.

Soil erosion is unfortunately becoming a serious problem in many places, especially along the eastern portion of the plateau, coupled with prolonged drier spells (Fig.4.04). Many of the soils have become useless, either because they contain too much or too little water. The soil, as a complicated mixture of rock particles, air, water humus and minute organisms including useful bacteria, changes with use and depth. The soil may fail to produce crops because of exhaustion, with the delicate natural balance of its composition becoming upset, which may also lead to enormous soil erosion (Hickman 1965).

Local settlers on the coast have long recognised that soil becomes poorer if it is constantly cultivated. African methods of cultivation have made allowances for such a problem. For instance, on the southern coast of Tanzania, the Makonde cultivators have been credited with having maintained the fertility of their land, which is naturally not very rich, by the practice of "thicket operation" (Gillman 1945; Hickman 1965). This is a form of rotation or changing of crops from year to year before allowing the land to lie fallow to develop thicket, after which it is cleared again to start a new circle.

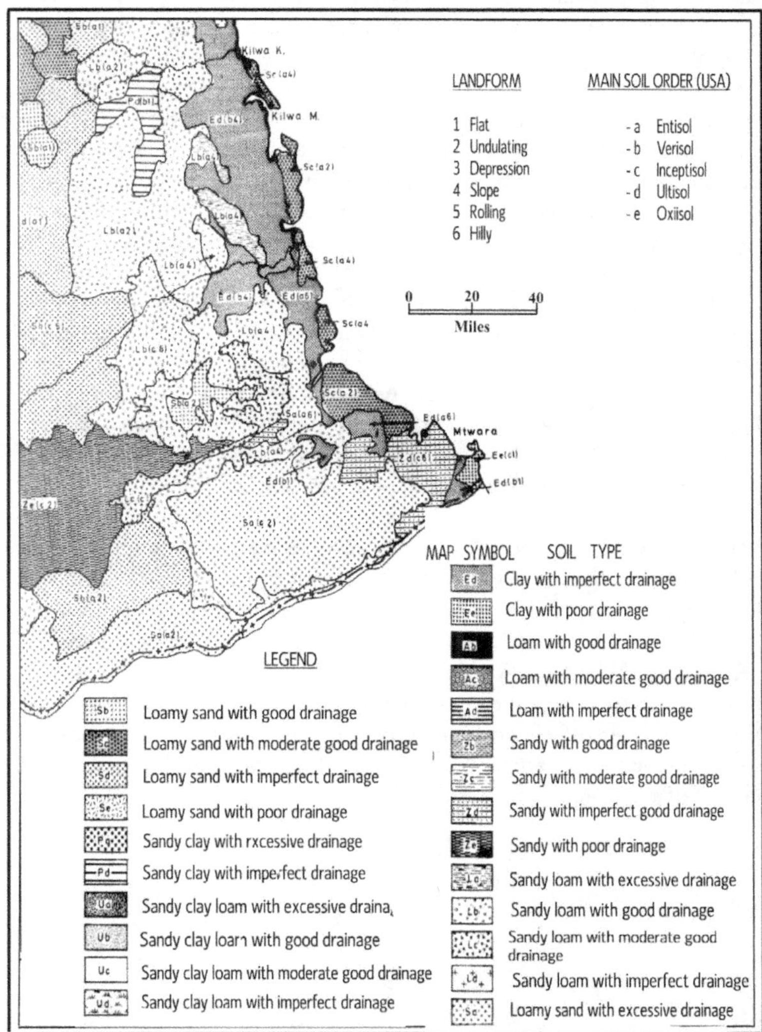

Fig. 4.10: Soils of southeast Tanzania, abstracted from S.A. Hathout, 1983

The beginning of soil erosion is almost always due to the soil losing its structure. That is, the humus colloids holding the crumbs together in the soil are used up or lost, and the particles fall apart. A soil can lose its structure through various ways, either resulting from human activities, or the wetting or drying of the soil. Among human activities the most common ones are over-cropping and over-grazing. However, although soil erosion is a natural process, it is made worse by man. In addition to the natural factors, the nature of the rainfall and the slope of some of the areas also play an important role. For example, rainfall can be too heavy at one moment (storms), especially after a prolonged delay.

In summary, all these parameters of geology, climate, fertility of the soils and vegetation cover have something to do with the establishment or abandonment of settlements. The interplay between environmental parameters and cultural factors tends to determine the patterning of a particular concentration of human and other populations in a particular favourable area.

It had been wrongly suggested that a dramatic change in the geomorphology of the region coupled with reduced rainfall and water levels were factors that probably caused settlements to shift with time, from its old preference on tops of plateaus towards the shores. In this work, a different interpretation can now be made after examining the environmental and cultural context of the sites. Although the distribution of settlements could partially be explained by the changing fertility of the soil, culture change has also played a considerable role towards the shift.

People of the area had been exploiting marine resources before, while processing them and living far from the water bodies. Currently, processing the shellfishes is mainly done on the shore and most of the settlements are also near the shore, although cultivation is still going on the some parts of the plateaus. Due to its small quantitative meat yield and low calorific value, shellfishes have been used as an alternative source of protein in an environment stressed for other sources of protein (Msemwa 1994). Alternating layers of shellfish horizons observed in some of the excavations during this work could be significantly suggesting fluctuations in climate of the area, hence periods of environmental stresses.

Chapter **5**

FIELD WORKS

The chapter on fieldwork focuses on with land surveys conducted on pre-defined data universe of this particular research, a narrow coastal strip of Lindi and Mtwara regions. It is divided into two main sections: surface survey and sub-surface survey that involved selective excavations as described in the following sections.

Surface Survey

This section is an account of land survey works conducted in several localities of the Southern coast of Tanzania. It is a systematic attempt to locate, identify and record the distribution of archaeological sites on the ground and in relation to their natural surrounding.

Archaeological survey is argued to be one of the most important components of archaeological investigations. It is concerned with the archaeological record of ancient settlement patterns, with ancient people's imprint on the land (Fagan 2005:176). Archaeological techniques for locating, studying and assessing archaeological sites vary from intrusive to non-intrusive techniques. Recently, archaeologists have paid attention to techniques of locating archaeological sites and assessing them without actual excavation. The latter normally changes the site forever. A study of entire landscapes by regional survey is now a major part of archaeological fieldwork (Renfrew & Bahn 1996). Strategies involved in this archaeological survey are discussed in the following sections.

Locating archaeological sites

It has been argued that some of the sites are quite evident, in that they are literally man-made structures, but others are far less easily located, perhaps displaying no more than a small scattering of stone tools, fractured stones, or potsherds or a patch of discoloured soil (Fagan 2005). Other sites have no traces of their existence or presence above the ground and may come to light only when the subsoil becomes exposed by erosion or other disturbances like tilling the land, digging or burrowing animals.

Surface finds of artefacts, ecofacts or other man-made materials and features have been used as the principal surface indicator of some archaeological sites. Various features, like vegetation, ploughing, and water and wind erosion as well as the nature of the artefacts themselves, are said to affect the visibility of these surface materials. Some remains may consist of thin scatters of artefacts or show up as dense concentrations or debris that stands out of the surrounding ground. Wind and water erosion or even burrowing animals have assisted archaeologists in revealing materials, which otherwise were buried beneath the subsoil, by bringing them to the surface (Fagan 2005).

Vegetation is also used as an indicator of a settlement site, as plants tend to grow more lushly in an area where the subsoil has been disturbed. Sometimes, specific types of trees or brushes are associated with archaeological sites. Baobab and some other nut-bearing trees in relation to the surrounding vegetation and landscape may suggest sites of human occupation.

Oral traditions have been particularly important in African archaeology. Memories passed on from generation to generation are important in many cultures, as stories, songs, and epics recount rules and important events in a community, as well as locating sites. Although they do not go back a long way in time, they show the existence of African unity in ancient times.

Soil marks, resulting from the exposure of distinctive soil types on the surface, were widely used to trace exhausted agricultural land. For instance, all of the EIW pottery came from distinctive reddish lateritic soil. The discovery of Pemba EIW sites in Mikindani was the result of such features, which Chami (my Supervisor) observed as he was flying over the area on his way back from Mozambique. A follow-up visit to the area revealed a truly rich EIW settlement.

Many more sites were also discovered by chance. Most of the archaeological sites are far less conspicuous, and without historical records to testify to their existence. Although chance and non-archaeological sources play an important part in site discovery, the greater part of the archaeological research involved the researchers in actively looking for the sites.

The Survey Strategies

Reconnaissance survey involved preliminary examinations of the survey area to identify major sites, to assess their potential and to establish tentative site distribution. It also involved the background research to acquire general information about the environment.

Intensive surveys followed reconnaissance. The survey area under discussion was traversed by automobile to a nearby predefined datum station before the footwork, which consumed most of the survey time and energy. On-foot survey enabled the researcher to scrutinise the topography and the relationship of human settlements to the landscape. However, it was not possible to map and cover every single point of the region, especially because of accessibility and resource factors.

The survey rather than being a general quantitative survey was guided by the research hypotheses and objectives. It ensured a wider range of samples in space and artefact diversity, than an examination of the whole region. Sites were approached using consistent, predetermined methods to ensure that all parts of the prescribed survey area were examined, instead of attempting to identify all the sites within a particular area. This approach leaves room for other interested researchers to work there and test the interpretations made by this work.

The sampling approach proposed by Mueller (1974) and Orton (2000) was employed for sampling many types of sites in each environmental zone in order to get a picture of past adaptations to different environmental conditions. It involved probability sampling, whereby small reliable samples of data were used to represent a much larger population (Renfew & Bahn 1996; Fagan 2005). Portions of a region, area, environmental zone, site or even artefacts recovered in this regard were selected to serve as representative sample of the larger area. The outcome of this approach depends on a very carefully drawn up research design and precisely predefined sample units.

The field survey reported here was organised and split into six sessions (Appendex IV). The first session consisted of eight weeks of survey and excavations in Mikindani areas (Mtwara) between June and August 2006. To make the fieldwork more efficient, some of the days involved both survey and excavations, depending on the weather and location. This system was also applied in other sessions, making it difficult to count the actual time spent solely on survey. The second session, which was of three weeks in February 2007, consisted of similar archaeological work in Rushungi and Kiswere areas (Lindi). Indeed, this was the most difficult phase that involved walking more than 23 km on rainy and muddy tracks before reaching the main road (see Fig, 5.01).

Fig. 5.01: The most difficult way back from the fieldwork after being caught by long rains at Rushungi site, on February 24, 2007.

The third session was five weeks of fieldwork in Mchinga and Mnang'ole areas (in Lindi) during July 2007. The session was followed by another four-week session of extra work at Mikindani and Kitere areas of Mtwara region. The last two sessions were for a small survey with more excavation work on two sites of Kilwa Island. During September 2007, two weeks of fieldwork concentrated on Nguruni areas. Another two weeks in November were spent doing similar archaeological work at Masakasa. However, following the previous comprehensive survey work on the island by Chami (2005), subsurface investigation was greatly needed rather than a repeated surface survey.

Surface Collections

It is argued that those artefacts and other archaeological finds discovered on the surface of a site are potentially vital sources of information about the people who once lived there. They provide preliminary information on the age, cultural association and types of activities represented on the site (Hester et al., 1997). However, they cannot assume the central role of any archaeological excavation.

Some archaeologists tend to distrust surface collections, arguing that such artefacts are easily destroyed on the surface or can be displaced from their original position by many agents. But, as a matter of fact, however deep archaeological deposits are found, they were once all on the surface for an unknown period of time and they might have been subjected to many of similar displacement and destructive processes as those suspected of the same today (Dunnel & Dancey, 1983). With increased emphasis on regional survey and settlement archaeology in recent years, many field workers have demonstrated that surface deposits can provide much information on artefact distribution and other phenomena found underground (Fagan 2005). Controlled surface collections that involved random sampling of mainly diagnostic artefacts were conducted, hoping to discover key finds that would identify the period of occupation, the activity area or cultural association represented at the sites. The operation was extended to cover the rest of the sites for highly diagnostic artefacts. Rigorous sampling techniques are said to be essential for obtaining even a minimum sample of finds at the individual site level, and that is where small excavations come in. Sometimes the surface finds may accurately reflect site content, although at other points, they may not. The distribution of these archaeological sites are usually plotted on large-scale topographic maps, though detailed mapping may follow for specific sites of interest.

In February 2006, archaeological reconnaissance was conducted as a preliminary examination of the survey area that covered a coastal stretch of about 250 km from Kilwa to Mtwara, to identify major sites, to assess their potential and to establish tentative site distribution. The first assumption was that most of the ancient coastal settlements relied on the sea for fishing, trading and long-range contacts based on sailing vessels. It is argued that the presence of a protected bay was important for the growth of a market town and port. The second assumption was that Kilwa became a big centre on the Southern coast, and involved in maritime trade long before the times of the Periplus documentation (Lacroix 1998). With the complex wind belts and ocean currents of the coast between the Southern coast and Madagascar, Kilwa had gained importance over other coastal settlements, linking the southern gold-bearing regions with the northern ports in a much shorter interval (Datoo, 1975). It was hence assumed that there should have been other centres on the area with a considerable population density that was suitable for safety ports, on some of the protected bays that appear on the Southern coast, which could have been influenced by Kilwa. A thorough search of the coast was expected to reveal ancient settlements that enjoyed the same coastal trade and resources as Kilwa.

Surveyed areas and results

The areas involved comprise of Mikindani-Kitere areas of Mtwara region, Rushungi-Kiswere bay, Mchinga-Mnang'ole, and Kilwa Island in Lindi region. As explained earlier, the areas form an enormous stretch of the coast of more than 250km, with poor infrastructure and limited resources (Fig.5.02). These areas were briefly surveyed at reconnaissance stage during the last week of February 2006 by the author and Chami, the Supervisor. At that stage, potential areas for intensive surveys were identified. Accessibility of the land is very difficult, especially when it rains; hence the logistics for assessing the areas were also sorted out during reconnaissance stage.

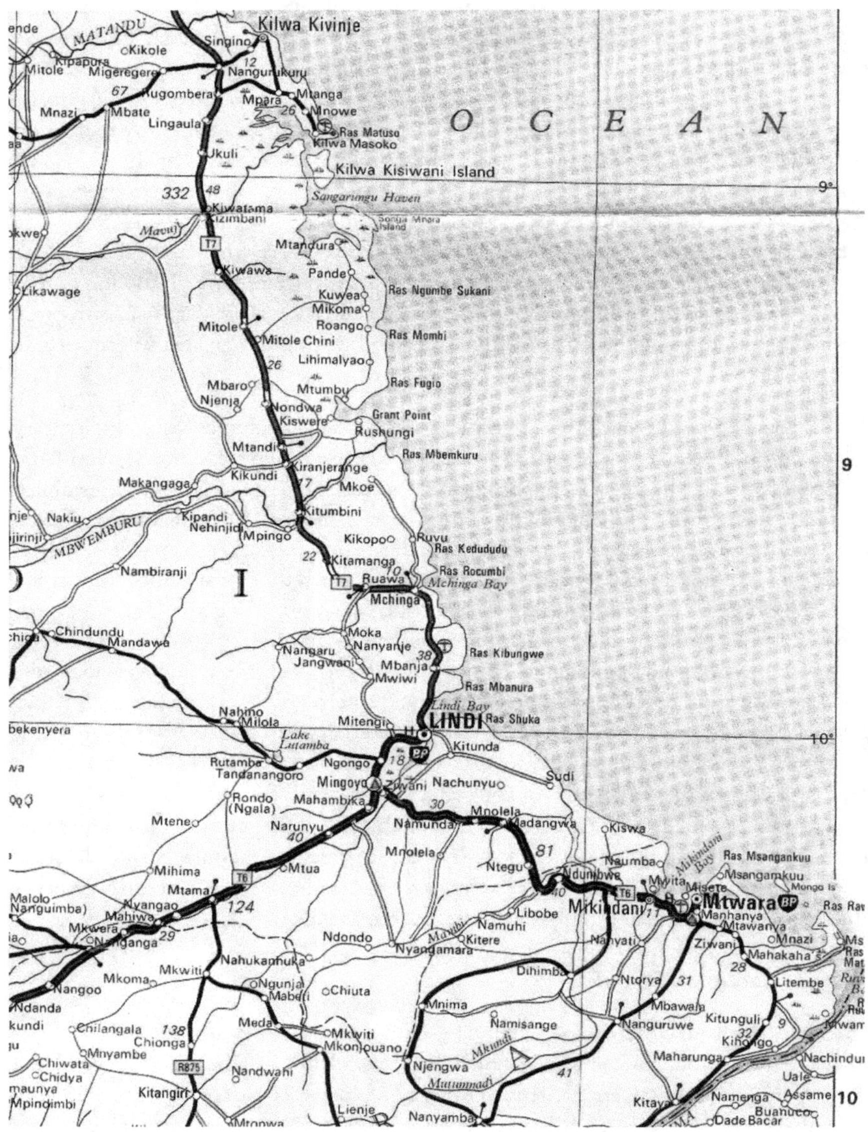

Fig. 5.02: A site location map.

Mikindani cluster

Mikindani area comprises a cluster of sites of different spans and complexity. A number of different sites were identified and recorded within an area of about 30 square kilometres. These are Mnaida (1-3), Nakatumbatu, Singino, Kabisela, Chikayowa and Pemba. The Mikindani town forms the core of modern settlements for these old settlement sites. The town lies roughly between 40° 6.2' – 7.2' E and 10° 16' – 17' S, 11 km before the modern regional capital of Mtwara (Fig. 5.03). The town is one of the 19th century Swahili towns that became an important port during the slave trade. It is on the shore of a well-protected bay with an extensive beach that was later developed by the Germans in the late 19th century as a regional administrative centre after the Lindi town.

Behind the town, the Makonde plateau rises to about 125m above the sea level but on the eastern and northern edge of the town the plateau extends up to the shore. The plateau is capped with thick and massive reddish soft sandstones and sands identified earlier as "Mikindani beds". The Lindi-Mtwara main road runs along the edge of the shoreline, making a natural northern border of the town. On the eastern edge of the town, close to the shore, is a big sand quarry for construction materials that exposes a wide and deep section of the Mikindani beds. Close to the sea the plateau has been dissected by erosion to form a series of hills and ridges that rise up to 122m above the sea level. The general geology of the plateau has already been discussed in Chapter Four.

During reconnaissance, a walk south to the Mnaida hills through this quarry (Fig. 5.03) revealed some very distinctive pottery not described in the pottery sequence of the Swahili coast (Chami 1998). The pottery looks old, thinned walls of less than half a centimetre, with a wide orifice but shallow in depth. It was immediately related to a PIW pottery, especially based on the cuneiform motifs of the decorations that filled the upper parts of the vessels without bordering (Fig. 5.33). Random filling with curving lines made up of rectangular or triangular points by rocker stamping and punctuation were common as seen in the analysis. This pottery, to be identified as "Mnaida tradition", showed some resemblance to the Nderit ware of the Rift valley, dated to between 7000 and 3000BP (Bower et al. 1977, Robertshaw 1990). A further search on the hilltop identified the richest locality (Mnaida 2) that obviously deserved further detailed work.

These early settlements were recorded on higher parts of the plateau that currently appears as a ridge, and away from any later deposits. To the western edge of the town, about 600m away, shell accumulations and pottery were observed on both sides of a ridge that runs westwards behind the old part of the town. The type of pottery includes that commonly known "Swahili ware" of the 2nd millennium (Chittick 1974;

1984; Horton 1984, Chami 1998), which is related to the stone towns' civilization. The rest of the area southwards to the Ruvuma basin was roughly assessed but without getting a better concentration of settlement sites.

The intensive survey programme during June-August 2006 exposed the richness of Mikindani as the largest site cluster as described below.

(1) **Mnaida hill**: This forms a ridge that rises on the southeast corner of Mikindani town. The fieldwork camp was located in the primary school bearing the same name located on the north-western slope of the hill. During survey, the hill was arbitrarily divided into 3 portions. Mnaida 2 (40° 07.2' E, 10° 16.9' S) comprised a stretch of about 500m of the ridge that runs southwards from the road, with the town cemetery at the foot of its northern slope. The PIW pottery that was later dated to the 3rd century BC was found on this ridge of about five hectares. Mnaida 3 (40° 07.1' E, 10° 17' S) runs west–east behind the school for about the same length to meet Mnaida 2 at almost what had been identified as Mnaida 1. The hill was noted as the best source of PIW sequence if thoroughly excavated.

(2). **Nakatumbatu**: (40° 07.3' E, 10° 17' S) is the higher ridge that runs north-south for about 3.5 km to the east of Mnaida, about 90m high at its north end but about 120m at its south end. An active valley with land slumps separates the ridge from the latter. The sand quarry mentioned before on the eastern edge of the town forms the northern end of Nakatumbatu ridge. Along the top of the ridge, especially the central part (Fig.5.03, loc. 8), extremely high concentrations of potsherds were observed covering a large area and the steep slope of the valley. None of the pieces that might count into thousands has any decorations. The concentration diminishes to the north and south of the ridge (Fig.5.03, loc. 9-10). The scatters extend to more than 200m toward the edges of the ridge on the eastern and western sides. There was a need to excavate to confirm their vertical distribution. It is now confirmed that people with Plain ware tradition occupied the ridge during the 9th century AD.

(3). **Singino** (40° 05.6' E, 10° 17.2' S): This is the highest edge of the plateau that rises steeply to 125m, located about 2 km southwest of the town. It becomes more flattened towards the south. On this ridge a number of stone artefacts, including Sangoan derived stone scrapers, were found scattered over an acre of now disturbed field (Fig. 5.03, loc. 11, 15). Pottery remains were similar to those of PIW of Mnaida 2, but without significant concentrations for excavation.

(4). **Kabisela**: is a village on the eastern side of Mirumba hill (40° 06.2' E, 10° 16' S), two kilometres from the town centre toward Lindi. It was noted during the reconnaissance survey that a road cut exposed an older settlement horizon on its eastern side that seems to have covered the entire hill before the establishment of a sisal plantation in the area (Fig.5.03, loc.12, 4). Settlements have now shifted to the lower terrace, especially to the foot of the southern slope of Mirumba Hill. Although activities seem to have continued on the site until the late Swahili phase with accumulations of marine shells from the exploitation of shellfish, eroding potsherds looked older, and were somewhat related to some of those included in Chittick's (1974) 'Kitchen ware'. Slightly flared rims with lip-indentations, together with single or double bands of vertical or oblique lines of incision or closely packed cardium shell impressions characterised most of the pottery. Sometimes incised bold lines bounded the bands to form a horizontal panel or festooned patterns.

The detailed work over the ridge indicated continued occupation ranging from TIW to the post-Swahili period. The eroding ceramics suggested mixed traditions with clearly Triangular Incised Ware, Plain Ware very similar to those from Nakatumbatu, Proto-Swahili as well as Swahili Ware. The size of the site could not be established because of restricted private use of the land. A search for undisturbed sequence was necessary to solve the puzzle. It is also likely that much of the site had been disturbed by sisal farming during the colonial time.

(5). ***Chikayowa*** (40° 05.6' E, 10° 16' S) is almost on the opposite side of Kabisela but on the other side of the road, about a kilometre inland towards the west. The Roman Catholic Church is also about a kilometre southeast of the site. A high concentration of Proto-Swahili pottery was seen in association with shell accumulations (Fig.5.03, loc.7). These types of pottery have not been described or reported from either the central and/or northern coast of the country. They are similar to the pottery reported to occur especially during Period 1b of Kilwa material mixed with a TIW pottery from underneath (Chittick 1966:7; 1974). Together they formed a large part of the so-called "Kitchen Ware" (ibid.). This Proto-Swahili tradition appears more distinctively at this site than formerly reported from Kabisela. Excavations could better show the history of the site.

(6). ***Pemba*** (40° 07.2' E, 10° 16' S) is a town on the northern side of the Mikindani bay, linked to the Naumbu sub-location. From Pemba it is 3 km by road to Naumbu and 2.5 km across the bay to Mikindani town but more than 6 km around the bay by land. It is much easier getting to Pemba using the waterway, but there is no formal ferrying service to Pemba, which is known to most Mikindani dwellers as an island. The survey around the ridges and flatness of the area showed varieties of Swahili wares. The last turn of the shore west of the town contained scatters of iron slag. A search in the higher parts of the plateau revealed a high concentration of EIW pottery.

The scatter extends to an area of about 100 x 25 metres (Fig. 5.03, loc.4, 6, 14). This was the first identified typical EIW site in Mikindani cluster. Only two pieces of thickened bevelled rims were found within Kabisela surface materials. At Mnaida 2, iron slag was found on the northern slope of the hill facing the town but rejected after realising that they came from Post-Swahili smith-work. Kabisela old railway sub-station used to be the source of iron scrapers for the smith-work. Slag-rich lateritic soil at Pemba and its associated distinctive pottery forms and designs is typical of EIW tradition. Only one potsherd showed incised triangular motif. From upper levels were shell accumulations of different weathered condition in association with Swahili Ware.

Extended survey around Mikindani and beyond could not locate a site similar to that of Pemba. This restricted occurrence of the EIW tradition in this area raised several questions without answers. Besides looking for the spatial distribution of the pottery, there was a need to establish their temporal range and relationships with the observed PIW tradition.

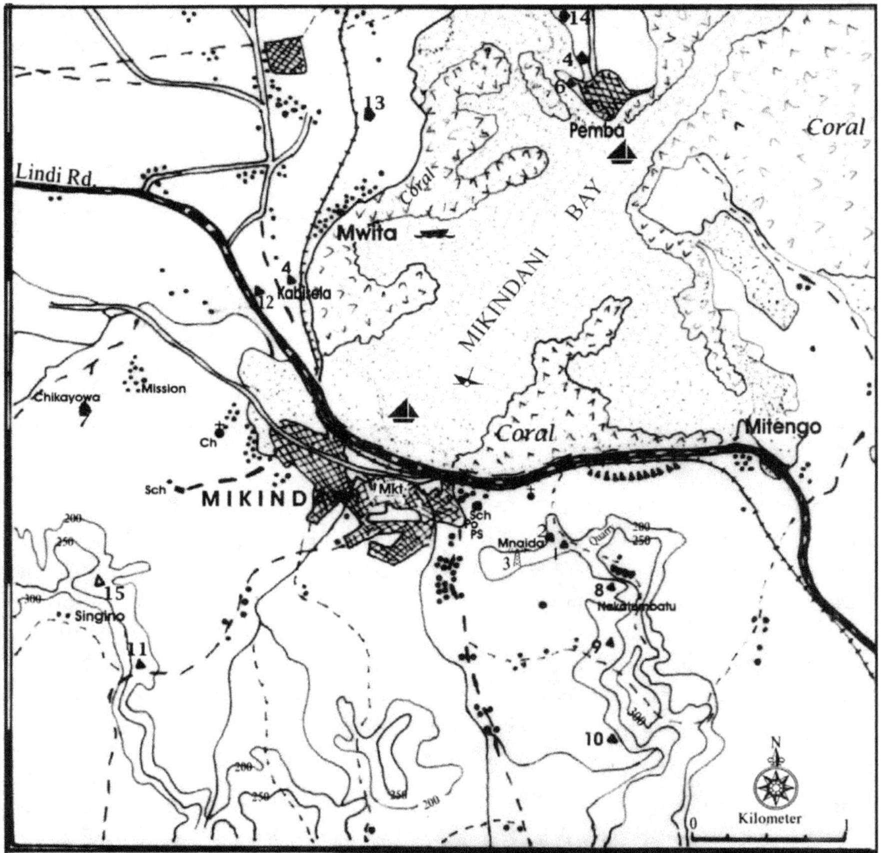

Fig. 5.03: A Map of Mikindani cluster showing the distribution of fourteen (14) archaeological sites.

Kitere area

Kitere was the first archaeological site to be reported on the Southern coast, 25 km inland (Fig. 5.02). The early finds of pottery from the site have been published elsewhere (Chami & Kwekason 2003: 70). They seemed to resemble Nderit ceramics of the Kenya/northeastern Tanzania Rift valley. Re-examination of the hill site during reconnaissance fostered the need for a detailed investigation that would possibly give a clue about the past relationships between the littoral and the hinterland. Continued research in August 2007 confirmed the previous observations that a PIW population had settled on the area.

Kitere (39° 46.7' E, 10° 21' S) is a densely populated village developed under the 1967 socialist policy. The village is close to the lake bearing that name, extending northeast along the Mambi River that emanates from the lake. The village is almost 22 km southeast of Mpapura village on the Mtwara - Lindi main road, and about 25 km from the nearest shore (Fig.5.02).

Important archaeological sites were found 3 km before the main village and one more kilometre up the hill that rises to about 180m above the sea level. The highest peak of the plateau is seen northwest of the hill site, which is 320 m above the sea level (Fig. 5.04). On the southeast side of the site, the plateau rises again from the river to about 305m. On the way from Kitere towards the east extends a high but flattened plateau for about 23 km to the Ndihimbe village on the Mbuo River.

Continued research showed scatters of fractured stones observed on the slopes and higher parts of the plateau. More convincing areas with concentrations of both pottery and lithic artefacts were the flattened top of the high plateau on the northwestern side of Mambi River, which is now dissected to form long ridges and hills. It was noted that some of the pottery belonged to what has been herein defined as Proto-Swahili ware. Its stratigraphic relationship could not be established through survey. Some of the fractured stones have their origin about a kilometre down the hill along the Mambi valley. Recent settlements are found down in the valley beside the modern village. Hence, the site required detailed sub-surface examination.

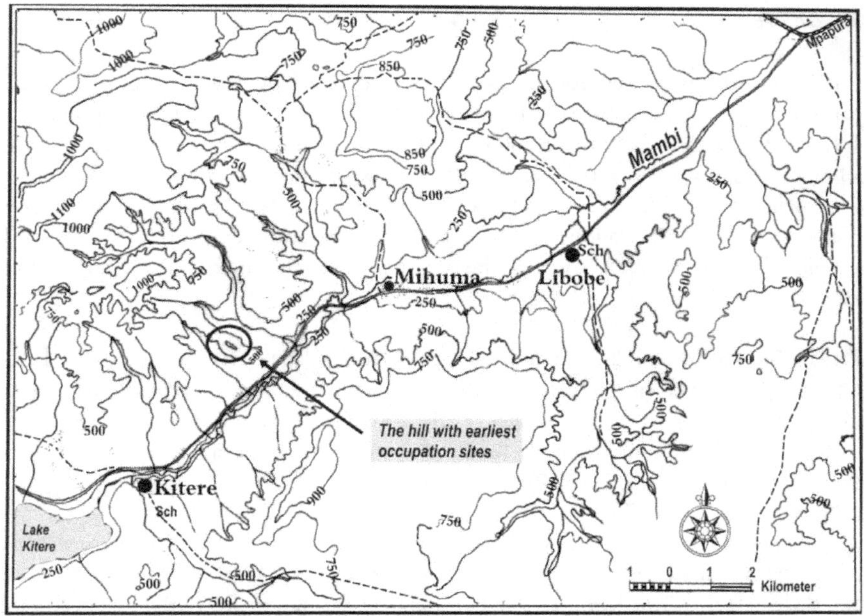

Fig. 5.04: A Map of Kitere showing the physical features and the site location

Mchinga bay

Mchinga is the next bay south after Kiswere. The western and southern parts of the bay are low and flat due to the Munimbira river valley. During the reconnaissance, the ridge behind Mchinga II on its northern side and eastern parts of Mchinga Secondary School showed scattering of pottery in now thicket-covered land. Most of this appeared to be of the late Swahili phase. Following the road to Ruvu bay and Mkoe, before descending into the creek forming Nondo bay, is a ridge capped with exhausted red sandy mantle with potsherds. In ascending from the creek to the flatness of the plateau is "Mkandiwata", an abandoned village site. Near a huge baobab tree of about 6 metres in diameter was a house foundation protruding above the ground on the road. In the reddish loam were potsherds of similar colour with thick walls of about a centimetre or more. One of the rim pieces had a bevelled lip. Although now in a thicket, the land is completely devoid of large/mature trees with the exception of three huge baobabs aligned to form a triangle, two of them along the road. The area was noted as a possible EIW site. Further north after Mnang'ole (Nangola), about a kilometre before the bridge on the Jalawa River towards Kilolambwani, similar patterns were observed along the road. Henceforth, the area was marked as worthy of intensive search for EIW settlements.

Mnang'ole village (39° 41' E, 9° 39.4' S) is located 15 km northeast from Kibiti-Lindi main road and about 3km from the nearest shore at Ruvu bay town (Fig.5.02). The village owns salt work project in the Nondo bay, 6 kilometres to the south. The road through Mchinga II to Kijiweni at Msungu bay on the Mbwemkuru river mouth now passes through the village.

During the intensive survey (July 2007) the earliest occupation sites were found in the southeast about 5 km from the village along the road, in what is called 'Mkandiwata' (39° 42.5' E, 9° 40.5' S). Mkandiwata, only 800m from the last bridge at Nondo creek up the plateau, is currently an abandoned area that has developed thicket after most of the inhabitants shifted north to Mnang'ole area (Fig.5.05). As it appears in most of the 1967 maps, initially the northward road used to pass through Ruvu town to Maloo (Mkoe) before shifting to pass through Mnang'ole and Kilolambwani. Ruvu is on the shore to the east, about 4 km from the road junction to Mnangole. It was once an important port, exporting sisal from Mkoe plantation. Currently, with a neglected road, it remains a port for local fishing although occasionally it exports marine shells for some ceramic factories in Dar es Salaam.

The survey work revealed a mixture of pottery traditions, and though dominated by what herein defined as Proto-Swahili Ware, it comprised some elements of EIW tradition. Some of the ceramics included fragments of conical crucibles with a gritty texture. There was an indication of iron slag, but not as quite common. The scattering of these artefacts appears at a length of about two kilometres within 50m from the road. Away from the road are mainly thickets that limit the visibility of any archaeological site. After ten days of survey and mapping, Mkandiwata remained among the sites that required intensive excavations.

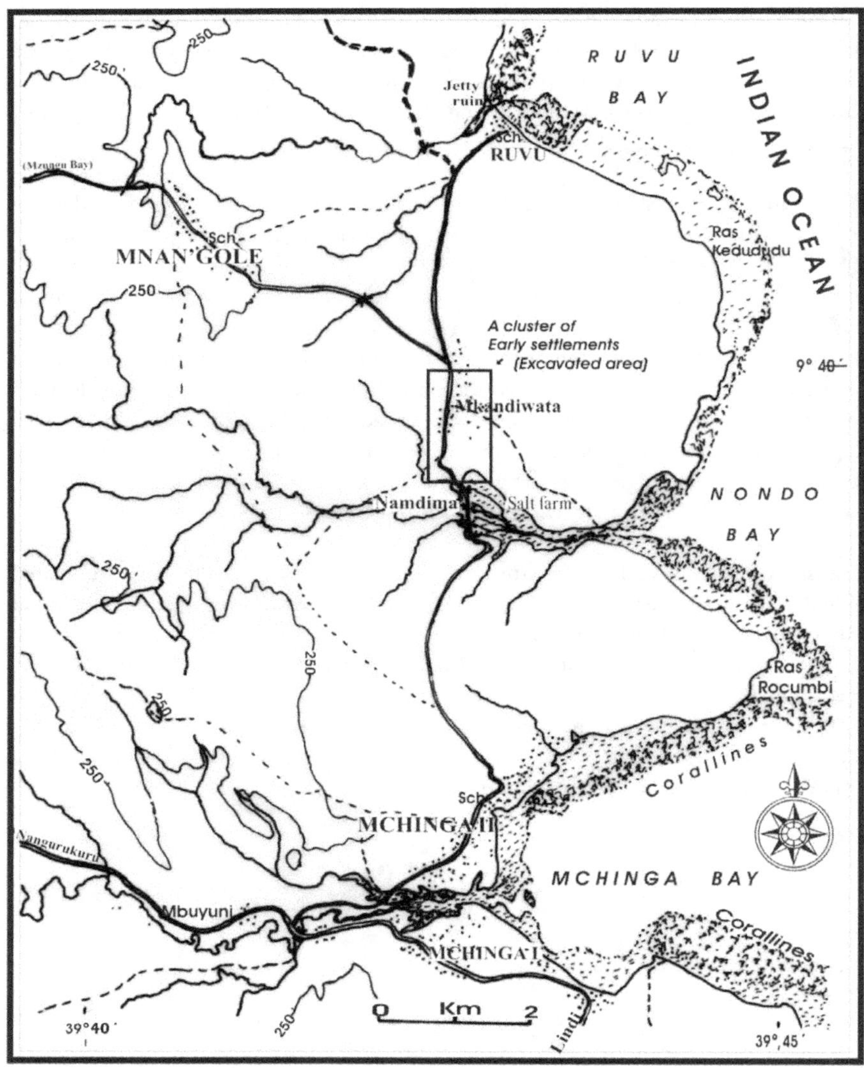

Fig.5.05: A Map of Mchinga-Mnangole sites cluster

Kiswere bay

The first well-protected bay south of Kilwa bay is Kiswere, with Mtumbu, Rushungi and Kiswere old towns (Fig. 5.02). Rushungi (39° 36' E, 9° 25.7' S) was probably the leading town on the haven. It is located 23 km on a very rough road from the Kibiti-

Lindi main road from a junction just before Mtandi (Fig.5.02); with a better port and an abandoned airstrip. During the reconnaissance survey, the first nice exposure of a site was found at Mkonya-ngonya hill. The vegetation is of a woodland type with thick undergrowth that made a systematic survey difficult.

A road cut had exposed an occupation horizon of about 30cm thick with marine shells and pottery accumulation at a length of about 40m. Most of the potsherds were decorated with either a narrow single or double band of oblique dotted lines, bordered by single lines of incision around the shoulder. Occasionally the band is filled with rocker-stamped impressions made by a multi-toothed object or shell-impressed. This was the first site to show this type of pottery that has not been described in any record on coastal archaeology.

The strong affinity of this pottery with those of the Rift Valley late-Narosura ware was so confusing. Hence, it was initially assumed to be a PIW site that could match the Narosura tradition of about 3000 – 2300 BP in the Rift Valley (Odner 1972, Bower & Nelson 1978). The survey of the hill and surroundings and a later visit to Rushungi town were enough justification for an early occupation and human activities on the bay.

An intensive survey was later conducted on the areas around the bay: Rushungi, Kiswere and Mtumbu-Mangesani locations. Rushungi is on the southern shore of the Kiswere bay. It had been exporting sisals from the estate that extended from the shore southwards for about 3.5 km to Mkonya-ngonya. Currently, the port is sometimes used locally to export limestone for cement factories. Kiswere areas have been important source of the limestone.

The road from Mtandi to the port passes though Mkonya-ngonya hill, a plateau that rises to about 120m above sea level. Older settlement sites are mostly on the northern gentle slopes of the hill (39° 36.5 E, 9° 27 S). The area has a single-component occupation of Proto-Swahili tradition with intensive exploitation of marine shellfish. Heaps of shell accumulations are common features on that hill. The shell mounds found on the sites are indubitably on a secondary man-made environment away from the shore depositional environment.

The only materials that have been found in association with the shells are pottery fragments and beads. As opposed to most modern shellfish collectors (Msemwa 1994), dumping of the shells was at the edge of living areas facing the slope far from the source areas on the shores. The roadway removed a larger part of the piles leaving undisturbed and flattened part of the shells on the eastern side of that road. Sisal farming might as well have destroyed a large part of the site on the northern side of the hill. (Fig.6.06). A thorough search of the area confirmed the absence of PIW tradition as previously assumed.

Kiswere is another town 5.5 km to the west of Rushungi on a shallow creek (Fig.5.06). On the western side of the town the hill is stony, made up of mainly crystalline and Nummulitic limestone deposits. The latter forms the base of the Mkonya-ngonya hill at Rushungi. The top of the plateau is thickly wooded. Down on the shore there is a German stone enclosure, constructed by freestanding blocks that are not cemented, probably a war-defence barrier. German bullet cases are scattered all over the field suggesting a battle field of 1905-07 Maji-Maji rebellion. Some potsherds were observed on the next hill across the creek on the southern side but without any convincing centre of concentration. They are all related to the Proto-Swahili ware.

Mtumbu is an old and abandoned medieval town on the bay north of Rushungi but west of Kissongo and Mangessani villages, still with some standing ruins (Fig.5.06). A two-storey building is still standing among big trees in a thicket. Plenty of German and Indian coins of the late 19th and early 20th were picked up on cultivated land near the shore. On the eastern side of Mtumbu are Mangessani and then Kissongo. Nothing of archaeological interest to this work was recovered in all of these areas with the exception of the higher plateaus that form the Mkonya-ngonya hill at Rushungi.

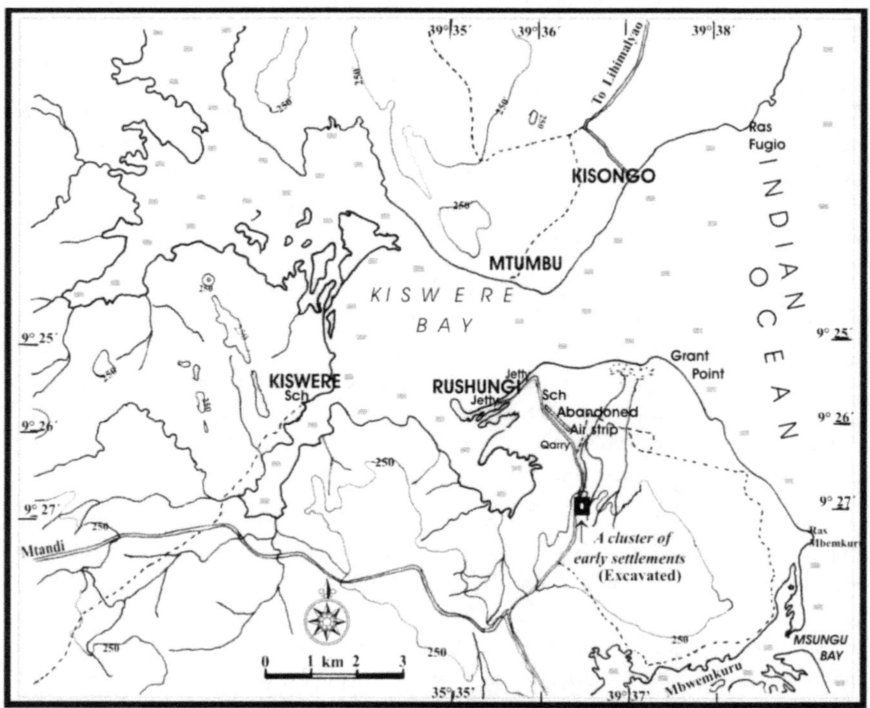

Fig 5.06: A Map of Rushungi-Kiswere sites cluster.

Kilwa Island

The island is quite known, perhaps more known than any other archaeological sites along the coast of East Africa. Excavations by Chittick on the island began in 1958 and continued up to the 1970s. Stone Age remains were reported from Nguruni, an area to the west of Hussun Kubwa (Isaac 1974: 254). The identifiable tools appeared to be of Sangoan type, comprising small hand axe-like points, irregular crude scrapers, etc. Some more similar artefacts were also found in the area to the east of Hussun Kubwa known as Masakasa. Hundreds of flakes, likely the relics of the manufacture of tools, were observed exposed on the side of the gully to the west of Hussun Kubwa. Glynn Isaac had visited the areas of Nguruni and Masakasa to examine the deposits and artefacts (ibid. 1974: 255-56). He suggested a Stone Age industry with Sangoan derived features, certainly related to an industry that occurs in the Southern province of Tanzania, currently split into two regions of Lindi and Mtwara, 60 km northwest of Lindi (i.e. upper-Mbwemkuru valley, towards the dinosaurs' sites of Tendaguru). Having seen a clear affinity with the Lupemban variant of Sangoan, he proposed the Kilwa and Lindi stone industries to be very interesting extensions of distribution and probably correlating with the belt of almost unbroken woodland that crosses southern Tanzania.

This research relied on the comprehensive archaeological survey of the island by Chami (2005) who mainly focused on the pre-Islamic archaeology of the area. Larger parts of Nguruni and Msakasa areas are still thicketed and without modern occupation (Fig.5.07). Stone Age industries were evident and typical bevelled EIW potsherds were also recorded. The commonly known Swahili ware is not well observed in this area, contrary to the central and western parts of the island. Quartz and sandstone pebble beds of estuarine bedded sands, conglomerates and shingles (Fig.4.03) are exposed on the slopes of gullies, in most cases sandwiched between grey, mottled, sandy loam of marine origin. The geology of these beds is more or less similar to Makonde beds, which comprise intercalation of reddish soft sandstone and grey marl that characterises the Makonde plateau.

The pebble bed source of the raw materials for stone working is exposed on the side of the gully to the east of Husuni Kubwa, about 15 metres above the present watermark. Currently in a number of areas around the northern margin of the island, gully erosion has cut back the coastal slopes, exposing red mantle and the marine formations, where stone artefacts were found among scattering of pebbles.

To the west of Masakasa and Nguruni is the old stone-town where most of the early excavations were concentrated. Nothing from the stone-town ruins is included in this work. The rest of the time, during the field session 5 and 6, was spent doing excavations in the mentioned two locations.

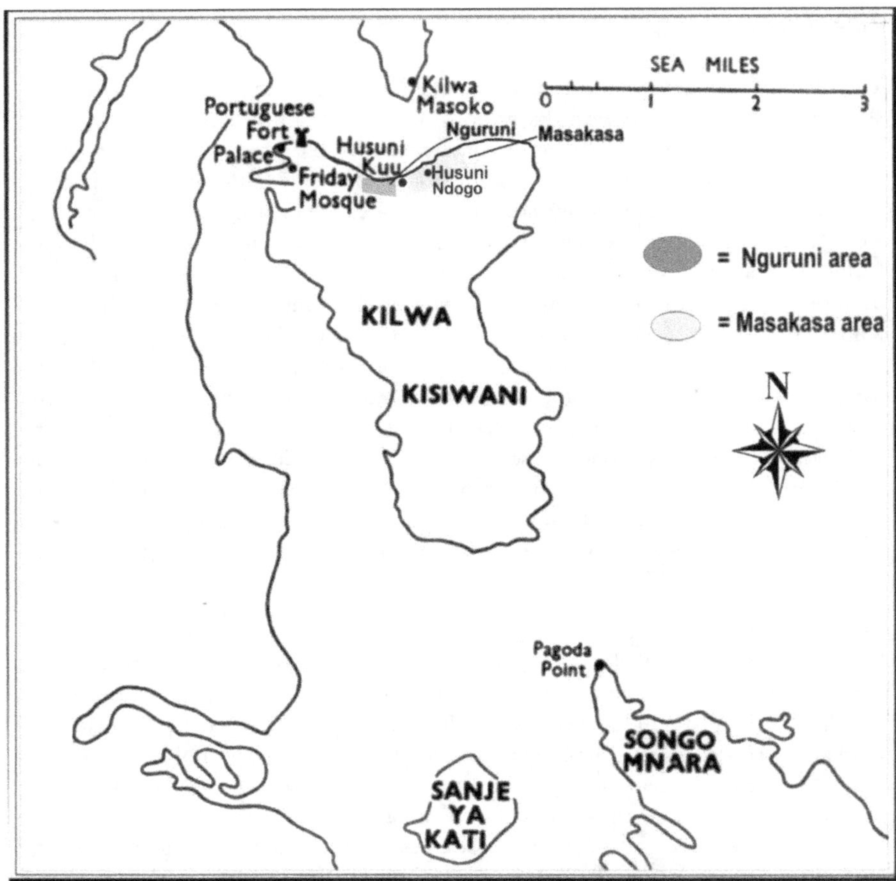

Fig. 5.07: A Map of Kilwa sites cluster

Upper Mbwemkuru

The road to the Tendaguru dinosaurs' site follows the woodland up the Mbwemkuru valley. This area has been known not only for the fossils remains but also for its scattering of well-finished Levalloisian flake tools (Smolla 1962, Isaac 1974). In recent times, Remigius Chami reported pottery, assumed to be one of the PIW variants (Chami & Chami 2001). A few stone tools similar to the reported above were also collected during this work. An attempt to relocate the pottery-bearing site reported by Chami (Chami & Chami 2001) was unsuccessful mainly due to inaccessibility because of the vegetation cover. The old road used by dinosaur researchers has forest cover and gully erosions. During the reconnaissance, there was a 15 kilometres drive inland from Mpingo village on the main road (south of Mbwemkuru River) to Nchinjidi and Mtarika villages before crossing the Mtshinyiri River that flows north to meet the Mbwemkuru. The land was fully vegetated and it was the beginning of the rainy season. The road had not been used for a long time. Coupled with heavy vegetation due to the last rainfall and gully erosions, it was difficult to find a way through the forest. A survey of the slopes of the eroding plateau revealed some stone tools of Middle Stone Age (Fig. 5.08) identified as Sangoan (Smolla 1962), which were collected. Isaac (1974) pointed to the distribution of this stone industry in larger parts of the woodland in southern Tanzania.

Archaeological assessment in the major construction work along the Kibiti–Lindi road was not so fruitful. Deep and wide fresh cuts along the road and cleared areas for construction gravels and substations were examined without recording any significant remains. These areas were then spared for a wider study of Pleistocene archaeology.

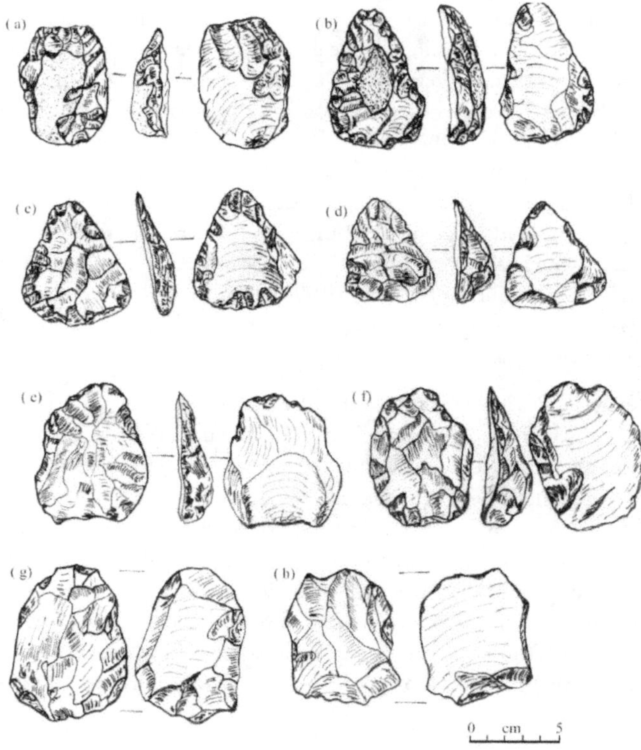

Fig 5.08: MSA tools from Mbwemburu valley near Nchinjidi area, Lindi: Sundry end and side scraper (a), unifacial point (b), bifacial point (c, d), convex end/side scraper (e-h).

69

Excavations

The focus of this section is on the gathered archaeological data from the sub-surface. Surveys and excavations are both major means of collecting evidence about the past, each entailing several methods of data collecting. It is argued that excavation retains the central role fieldwork because it yields the most reliable evidence for features not detected by other techniques, and helps to construct a detailed picture of features already identified by other means (Renfrew & Bahn 1996:96). The most reliable samples for temporal relationships are also collected through careful excavations. The details of the sites have been presented in section on Survey areas and results of this chapter, P. 61.

Subsurface Techniques

Methods for sub-surface investigations range from simple ones, to those requiring expensive equipments. Most of these techniques are said to provide limited coverage, and are time and resource consuming. The most direct and simple methods are augering, coring and shovel testing. While an auger is a large drill that uses human or sometimes powered motor, corers are hollow tubes driven into the ground that when removed, yield a narrow column or core of matrix, providing a quick and relatively inexpensive cross-section of the sub-surface. Coring is most useful in areas with a stratigraphic depth of more than ten metres (Fagan 2005: 196; Renfrew & Bahn 1996:85). This method has been employed recently in some of the East African sites along the coast and islands (e.g. Juma & Löfgren 1992; Sinclair et al., 1992; Radimilahy 1992). Augering, which is more suitable for a shallow stratigraphy, has been employed on the central coast of Tanzania (Chami 1994). However, Renfrew argues that there is always a risk of damaging fragile artefacts or features by this method, besides its limited coverage (Renfrew & Bahn 1996).

Magnetometer, resistivity detector, and pulse radar are some of the sophisticated instruments, which have been applied in sub-surface archaeological surveys. Concentration of archaeological features or materials tend to create anomalies in the intensity of the magnetic field or resistance to electric current. Electromagnetic waves also tend to be reflected as echoes by sub-surface discontinuities or solid objects. The recording of these anomalies and echoes has successfully indicated the presence of buried features in particular (Fagan 2005:194-95). But as mentioned earlier, the instruments are costly in terms of labour, finance and time. Magnetometer was applied shortly on Limbo site on the central coast of Tanzania while searching for buried furnaces (Chami 1992). However, none of these techniques could be afforded during these investigations.

Excavation is argued to be the principle means by which archaeologists gather three-dimensional data from beneath the ground. It is the only way to verify features and learn the characteristics of sub-surface data even after application of expensive remote-sensing techniques. There are general methodologies for excavation, but the appropriate one varies from site to site and from moment to moment as an excavation proceeds. According to Renfrew & Bahn (1996: 100), all excavation methods need to be adapted to the research question in hand and the nature of the site. They must also balance destruction of the site against the needs.

The Excavation method

In this research, vertical excavations formed the central focus (see Renfrew & Bahn 1996:101-102). Trenches were systematically opened to uncover the chronological sequences of cultural material and hence cultural dynamics of the region. Occasionally the trenches were extended on either side beyond the trench boundaries to affirmatively clarify features like a hearth; a series of stones or anomalies and concentrations of remains encountered near trench walls. All the excavations were dug to arbitrary levels, whereby the soil was removed to a standard sized arbitrary depth that may vary in size depending on the nature of the site. In all the cases of this work, the arbitrary level thickness was confined to 10 cm. This approach is used when there is little discernible stratigraphy or variation in the occupational layers (Fagan 2005: 218), by which each level was screened carefully through 5 mm wire mesh to recover artefacts, animal bones and any other small finds of relevance to the research questions. This technique is argued to be the best option when natural layers are not visible or indeterminable based on colour variations. Normal tools for these excavations were mainly sharpened masons' pointed trowels, geological pick-hammers, fine brushes and occasionally picks and shovels on especially hard clayish soil.

A careful selection for sites to be excavated was carried out in all of the surveyed areas. For Kilwa Island, the excavation was done away from the stone town where intensive excavations had been conducted in the earlier times (e.g. Chittick 1966; 1974). The new areas include the north-eastern shores of Nguruni and Masakasa locations that are currently unoccupied. Indeed, the area provides the oldest evidence of human activities in the island. These excavations were carried out archaeologically to reveal the vertical distribution as well as spatial relationships of artefacts within and between sites. Analyses were designed to provide information for the archaeological record. Tests were also done through absolute dating that involves carefully collecting charcoal samples from excavations, which were analysed in Angström laboratory in Uppsalla, Sweden. The final examination of the basic questions formulated at the beginning of the project was conducted after the fieldwork to compare the new finds and data including regional comparison with data from other related sites from East Africa.

In the following sections are the descriptions of the excavation results from all the sites considered potential for a sub-surface examinations during the survey operation.

Mikindani cluster

Mikindani, especially at Mnaida Hill, comprises the most convincing evidence for the earliest settlements in Mtwara region. The observed potteries on this hill were unique in style and design, different from the rest of the discovered sites in Mikindani and other areas.

The hill was excavated in two field sessions. The eight weeks of the field season between June and August 2006 concentrated on the Mikindani area in general. An area of 45.6m² (square metres) was excavated in the form of 10 isolated trenches in five different areas around Mikindani. Five of the trenches were on the Mnaida hill, two at Pemba and one each in Nakatumbatu, Kabisela and at Chikayowa areas.

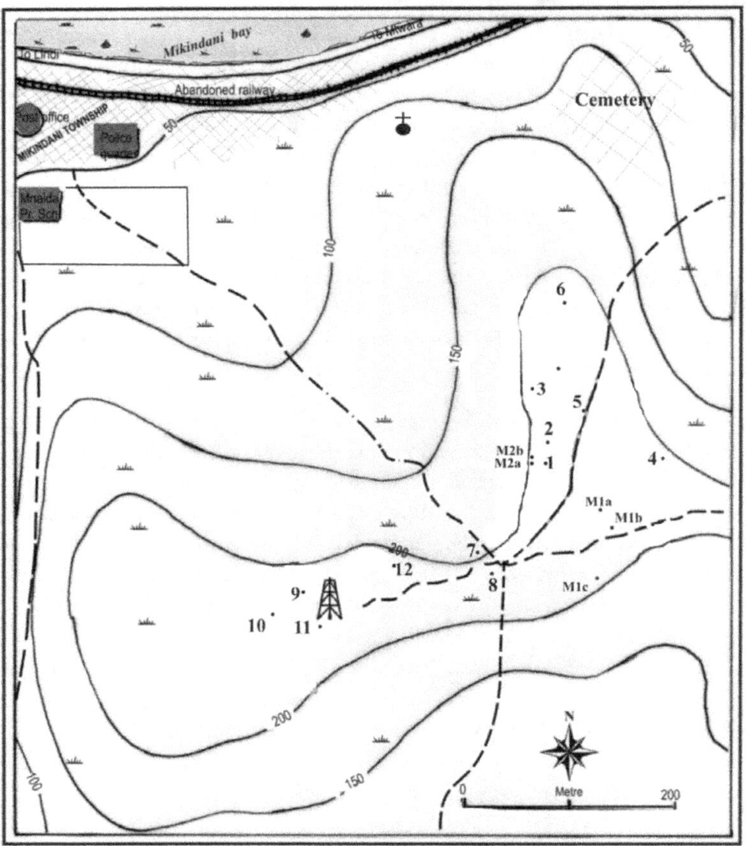

Fig. 5.09: A map of Mnaida hill indicating the distribution of trenches by numbers. The mobile telecommunication tower near trench number 11 is a landmark at Mnaida 3.

72

During 2007 excavations 6m^2 were excavated in two trenches on top of the hill site in Kitere area (Fig. 5.19). Twelve other trenches covering an area of 63.5m^2 were also excavated at the top of Mnaida hill that forms an **L**-shape with Mnaida 1 at the centre (Fig.5.09). Together, the excavated area around Mikindani itself during the two seasons count to 117.1m^2, 50 % of the entire excavations of this research (Fig.5.28). In the followings subsections are specific details of the Mikindani localities.

Mnaida hill

Mnaida hill is composed of generally shallow cultural horizons of less than 60 cm. On several areas of the hill the lowest pottery horizon has been exposed through rain-wash of the overlying soils (Fig.5.10, 5.11). The remaining soil is exhausted and sandy, and without recent deposits on top. Variation in the depth or level enriched with this early pottery is mainly the function of the current surface altitude rather than absolute depositional differences.

At Mnaida 1, three trenches of 2x1m each were excavated during the first field session in 2006 (M 1a-c on Fig.5.09). The soils are mostly unconsolidated dry reddish silty sands with less than 4-centimetres of humic cap. There is very insignificant colour variation down the Mikindani sands that become more reddish gradually below the thin upper cap. Based on this, the excavations were at arbitrary levels of 10cm each. Cultural materials were found within the upper 30cm and below them were indeterminable thick sterile sands. Pottery of probably the same tradition (n=28pcs) and more than 20 highly weathered marine shells were recovered. A few pieces of frankincense were also recovered from especially level 3 (20-30cm). The pottery is considered to be part of the PIW Mnaida tradition (Fig.5.33-5.34).

At Mnaida 2, two trenches of 4x3m (M 2a) and 2x3m (M 2b) were excavated during the first session in 2006 (Fig.5.09). The soil is more or less the same as the one at Mnaida 1, more silty and brownish at the top but more sandy and reddish after 3-4cm. Only the first three levels contained artefacts, i.e. pottery (n=186), marine shells (~ 90) and frankincense. Twenty five potsherds with decorations were identified as part of Mnaida tradition that has been dated to the last millennium BC. The fourth level (30-40cm) had nothing except three small, undecorated sherds. During 2007, five trenches ranging from 4- 9m^2 in size were opened as shown numerically on the Mikindani map (Fig: 5.09). 630 potsherds were collected from the upper six levels, of which 59 are decorated. Among the pottery, only 11 pieces came from the fifth (40-50cm) and four from the sixth levels but none of them have decorations. Four fractured stones were collected from the fourth and fifth levels. Frankincense was also collected and a few marine shells at different levels. Below the sixth level (50-60cm) the soil is sterile and brick red. A quick observation of the shape and decorations suggested a single component tradition, here called Mnaida tradition except for trenches 2 and 4. In these two trenches the first two levels provided Proto-Swahili pottery.

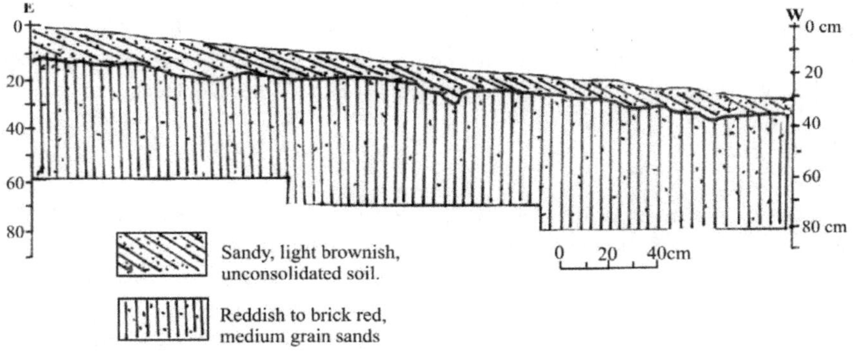

Fig. 5.10: Southern profile of trench M2 at Mnaida 2

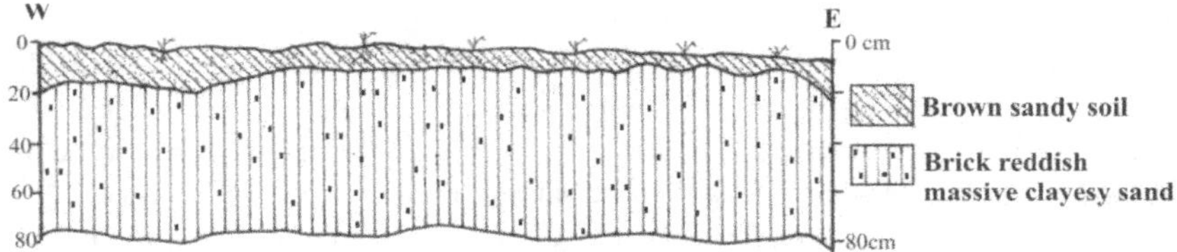

Fig.5.11: Northern profile of trench 6 at Mnaida 2

On the other side of the hill (Mnaida 3) another six trenches ranging in size from 2m² (2x1m) to 6m² were opened, depending on the surface concentration and location. More than 1200 potsherds were collected, 97 of them with decorations. A few marine shells (n=18) and frankincense pieces were also among the assemblages. From trenches 7, 8, 9 and 11, PIW potteries (Fig. 5.33) were collected from the upper three levels. Trench 10 was the richest in terms of pottery (n=560), of which 76 were decorated. All the potsherds from levels 1-6 appeared to be of Proto-Swahili tradition. Pottery recovered from levels 7and 8 had no decoration and below was sterile soil. Four beads were also recovered from levels 2 and 3. One sherd from level 2 (10-20cm) has decorations on both sides. In a similar way, trench 12 provided Proto-Swahili from all the levels. One sherd from level 4 (30-40cm) was decorated on both sides. The details of the trenches are provided in Appendex I (pg.281).

Nakatumbatu ridge

Nakatumbatu, which is on the southeast of Mnaida Hill with a stream valley separating the two, turned out to be the best area with the Plain Ware tradition. The central parts of the ridge (40° 07.3' E, 10° 17' S) showed the highest pottery concentration, most of which had no decorations. They were scattered in a similar manner for about 100m without mixing with any other known traditions. In 2006, a small trench (1x1.6 m) was opened and the richest horizon was found between 20 and 30cm (Fig.5.12). The upper levels had a similar material. In this level 3 (20-30cm) more than 640 potsherds in a weathered condition and rough surfaces were recovered (Fig.5.35). Only two pieces, probably of the same vessel from the upper second level, appeared with very fine, closely packed short oblique incisions on the shoulder.

During 2007, a trench of 1x2m was excavated that provided only one among many sherds with decoration from the first level. Below level 3 the soil is sterile. Hence, similar to what was observed on the surface, the excavations confirmed the site to be a single-component site, occupied only during the Plain Ware tradition. A well-preserved charcoal sample from the lower level 3 in association with the potsherds was recovered for absolute dating (see results in Table 6.14).From the analysis, the tradition is shown to date to the 9th century AD.

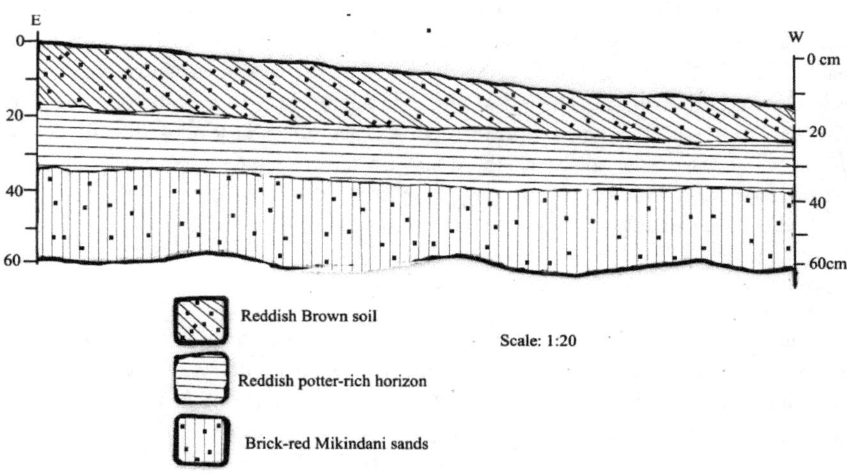

Fig. 5.12: Southern profile of trench M7 at Nakatumbatu ridge

Mirumba hill

Kabisela site on the Mirumba hill (Fig. 5.03) was excavated during 2006. The site extends into a fenced private plantation. Through erosion, the south-eastern extension of the site has crumbled into the sea and towards the lower terrace with modern village. Excavations were conducted near a thick pile of bivalves and gastropods shells from shellfish harvesting in association with pottery at the edge of a gully. Two trenches of 2x3m and 1x3m were excavated, revealing a continued occupation of the site from probably late Ironworking to the late Swahili ware period.

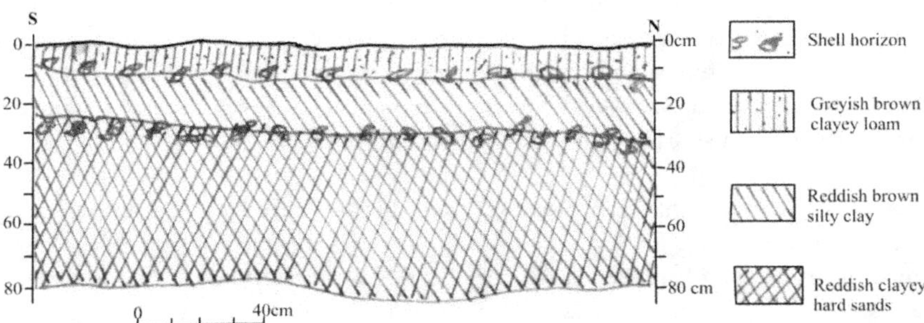

Fig. 5.13: Western profile of trench M3 at Kabisela

The upper three levels are of fine loamy soil, consisting of a mixture of Post-Swahili and Swahili ware. From level 4 (30-40cm) the soil becomes more clayish and hard. In level 4, mainly Proto-Swahili ware was recovered in association with three Indian red beads. Level 5 provided only a few sherds (n=5), highly weathered and have a gritty texture without decoration. They were similar in attributes to that from Nakatumbatu (Fig. 5.35). Some of this type of pottery came from the lower level 4, directly below Proto-Swahili ware. Towards level 6 (50-60cm) potteries become fewer, relatively thick and in association with some Iron slag. They seemed to belong to a TIW tradition that had been collected from a gully three metres from the trench (Fig.5.36).

Chikayowa

At Chikayowa (Fig.5.03), an area of 12m² was excavated in three trenches during the first and fourth field sessions. The achieved lowest arbitrary level with cultural material is 70cm (Fig.5.14). In the first session (2006), a trench of 2x3m was opened. Two more trenches of 2x2m and 1x2m were opened during the 2007 session. The soil is dark-brown clayish loam on top that becomes lighter, fine and hard further down.

The first level and upper second level provided mainly a Swahili ware. Carinated forms and Chinese ceramics were common here. From lower level 2 (10-20cm) downwards, the characteristic pottery was mainly Proto-Swahili ware (5.37-5.38). Assorted beads (n=18) and bones, the identifiable pieces of which included dik-dik horn cores and teeth, indicating wild species, were also collected (Fig.5.30). Together with the wild species, fish remains, including unidentified large species, were collected, some of them having been burned. At 65cm, in level 7, the soil became of compact hard sterile clay.

Chikayowa is the site that provided most of the collected bones from food remains distributed randomly from levels 3-7 in association with Proto-Swahili ware (Table5.01, Fig.5.29, 5.31). More than 3700 pieces of pottery were recovered from those excavations at Chikayowa, 397 of them with decorations.

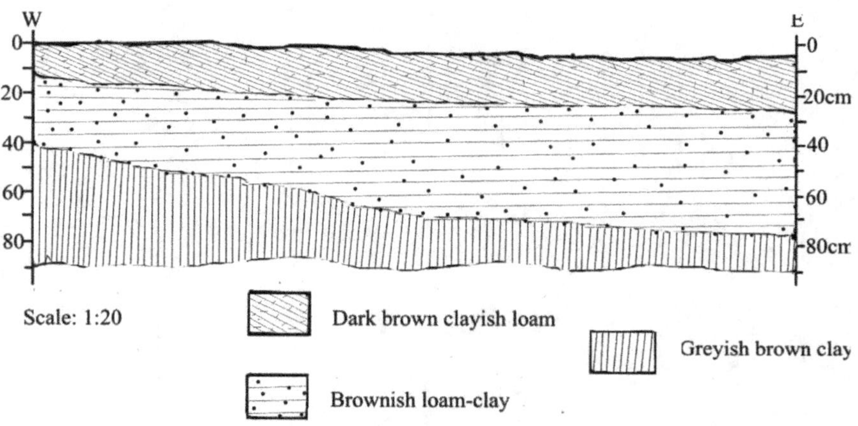

Fig. 5.14: Northern profile of trench M6 at Chikayowa

Pemba Peninsular

Pemba has been mentioned as an important and unique site in Mikindani. It provided the richest concentration of an EIW pottery and iron slag on the surface during the 2006 survey. Pottery that has been identified elsewhere as Proto-Swahili ware was not found at all in Pemba. The elevated parts of the area were capped with laterite-rich soil, possibly a potential source of the raw material for iron smelting in the early times.

Two trenches of 3x3m and 1x2m were excavated in a cashew nuts field (Fig.5.15) namely M4 and M5 respectively. The soil consist of unconsolidated reddish Mikindani beds of almost uniform colour and fine texture (see Fig.5.16). The upper part was slightly light brown, probably due to leaching. The tone increased gradually downwards to become brick red.

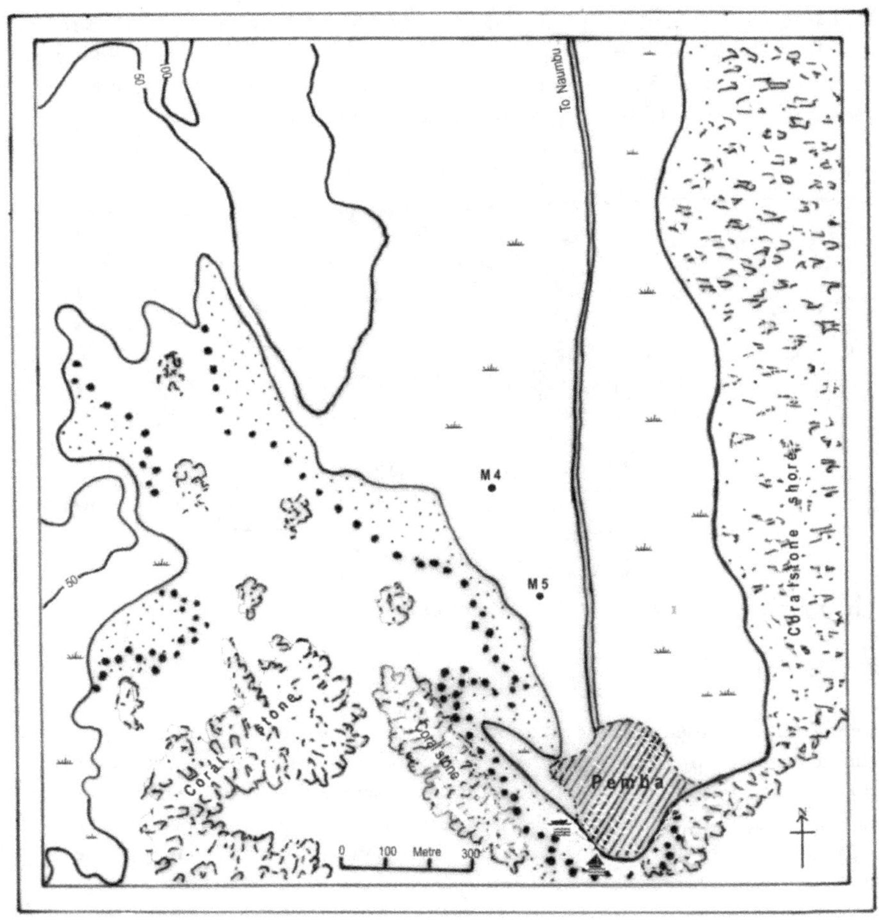

Fig.5.15: Pemba peninsular and the town (stripped) in relation to excavation localities (M4&M5)

78

Fig. 5.16: Excavation works through the Mikindani beds at Pemba site, Mikindani

Stratigraphy of Pemba is much deeper than the rest of the Mikindani sequences (Fig.5.17). The first level provided pottery of the Swahili ware, Chinese porcelains and marine shells. More shell accumulations were witnessed on the western slope, 5 metres from the trench M 4. The lower level 2 of both trenches showed pottery probably of a TIW tradition of which two sherds have an incised triangular motif. Indisputable EIW pottery in relatively large pieces and with decorations was recovered from the third level (20-30cm) down to level 8 (80cm). Iron slag was common throughout these levels (Fig.5.39-5.40).

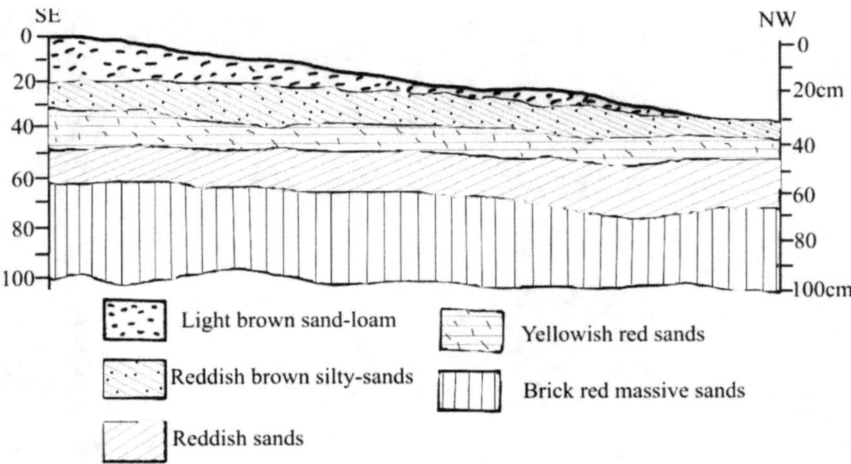

Fig. 5.17: SE-NW profile of trench M4 at Pemba peninsular

These finds make Pemba site the richest known EIW site in Mtwara region. Imported ceramics appeared in the upper level 3 (~20cm) or more precisely lower level 2. Pieces of charcoal for dating were also collected from different EIW levels that provided a date range of the 4th-5th century AD (Chapter Six).

Kitere hill

Kitere area was excavated during the 2007 field season. Two trenches of 2x2m and 1x2m were excavated down to a depth of 70cm where sterile hard clayish soil was encountered (Fig.5.18, 5.19). The first trench had been disturbed by a house foundation that went down to the bottom of level 4 (40cm). The soils became hard sandy clay from the fourth level onwards. A number of lithic artefacts were also collected in the fifth level in association with a few pieces of pottery possibly of PIW (Fig.5.41). Level 6 of this trench (50-60cm) provided only lithic artefacts. In the other trench, Proto-Swahili ware extended from levels 1 to 3 (Fig.5.42), and a PIW pottery in levels 5 and 6. Lithics appeared with pottery from the third to the fifth levels. The sixth and seventh levels provided only lithics. Although pottery appeared up to a depth of 50cm, those with decorations were recovered from above 40cm depth. Below 35cm deep, fractured stones pieces appeared with pottery. Some of these fractured pieces were of chert-variety of rock whose natural occurrence was about a kilometre down the hill.

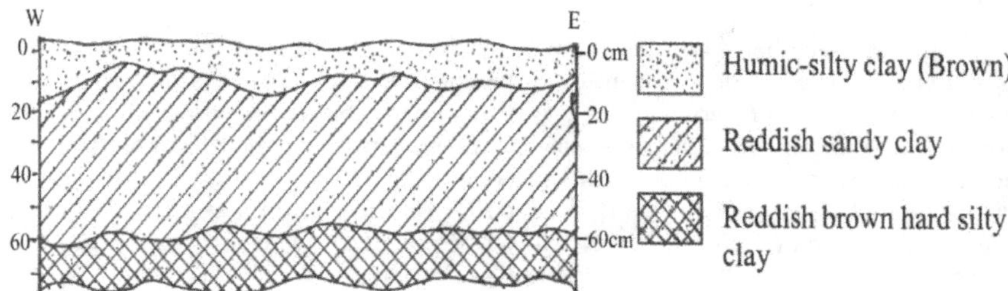

Fig. 5.18: Northern profile of trench 1 at Kitere

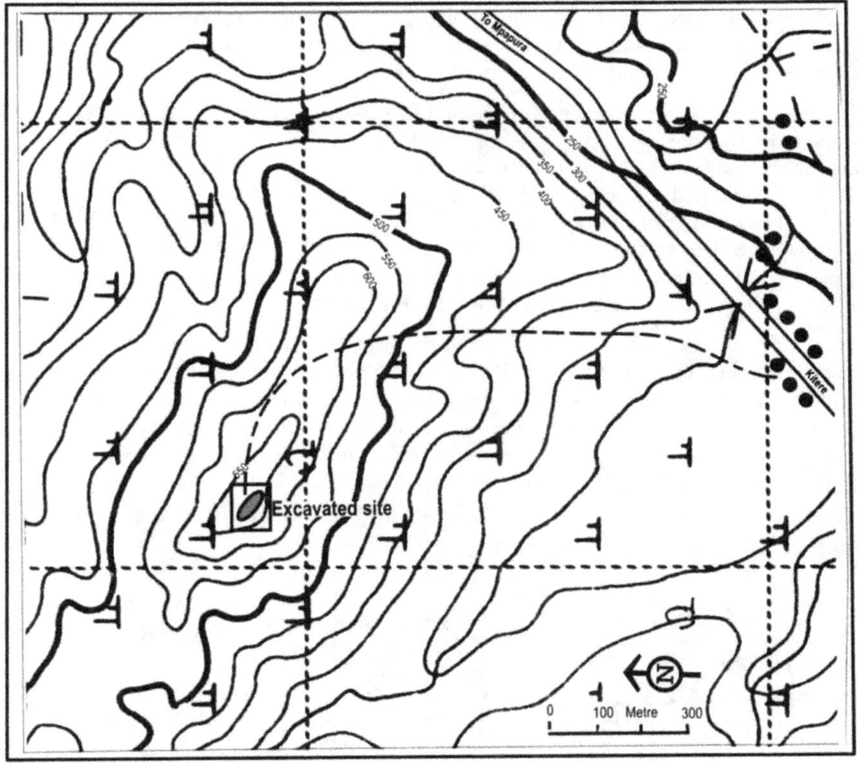

Fig5.19: Kitere hill Map showing the site and the excavation location

Mkandiwata terrace

The higher terrace between the Mchinga and Mnangole areas as one ascends from Nondo bay showed evidence of an EIW occupation during the reconnaissance. The area between two huge baobab trees closely aligning with the roadway at Mkandiwata showed the richest sites during the survey. The settlement seemed to have extended across both sides of the road. At this place, an area of 35.5m^2 was excavated in 12 distinct trenches (Fig.5.21). This is almost 15% of the entire excavated area during the entire research (Fig.5.28). More than 6,000 potsherds were recovered from this excavation alone and 10% of the amount had decorations.

From trenches 1 to 4, recovered pottery was mainly of Proto-Swahili that extended, from the first level down to levels 4 or 5. There was a bevelled rim in level 3 (20-30cm) of the first three trenches and level 2 in trench 4. The rest of the pottery is typical Proto-Swahili ware. Trenches 6 and 9 also provided only a Proto-Swahili sequence from the top to the sterile level (Fig.5.45-5.46). The most interesting sequence is that of trenches 5, 5E, 8, and to some extent trench 7.

In trench 5, the upper four levels contained Proto-Swahili ware. The first bevelled rim came from the lower part of level 4 or possibly part of the upper level 5. There was a drastic change from level 4 to level 5, with relatively very few sherds, the disappearance of decoration and introduction of thick pottery with bevels. Proto-Swahili pottery ends at level 4. A light green glaze ware of white lime body, possibly Sassanid pottery, was recovered in association with EIW ware in this level 5. Levels 6-8 contained typical EIW pottery (Fig.5.43-5.44), some iron slag and copper-working crucibles. The latter were also found within Proto-Swahili levels.

Trench 5E showed some Swahili ware in the first level. Levels 3 and 4, though with some iron slag, contained typical Proto-Swahili ware. From levels 4-6 typical EIW pottery was evident that was associated with chipped stones/artefacts in levels 5 and 6. The last two levels after these contained smaller fragments of pottery and flaked stone pieces without diagnostic pieces. In trench number 7, the first level, a Swahili ware was found followed by two levels with Proto-Swahili ware. Only one potsherd was recovered from the fourth level (30-40cm) after which the soil is sterile. This potsherd has a bevelled rim suggesting an EIW horizon.

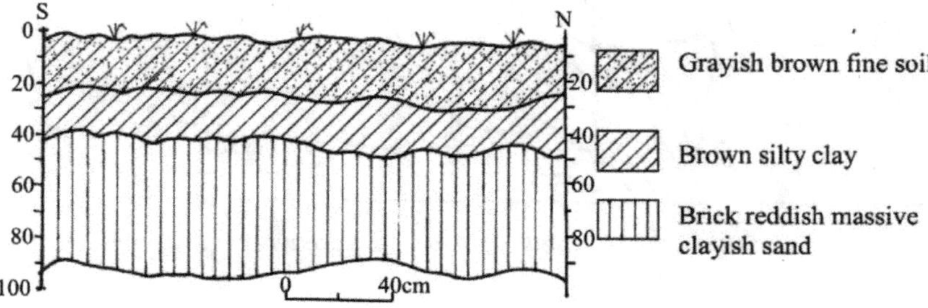

Fig.5.20: Western profile of trench 5E at Mnangole

Trench number 8 provided pottery of Swahili ware in the first level; Proto-Swahili ware in levels 2 and 3, and some iron slag in level 3. From level 4 (30-40cm) were relatively big potsherds of typical EIW pottery. Level 5 was the last level with cultural material but no diagnostic potsherd. Again in trench 9, potsherds of EIW pottery started to appear in level 4, below Proto-Swahili ware, with a relatively high concentration of iron slag but there was no diagnostic piece among the recovered potsherds (n=44). This continued to level 5 before the sterile soil in level 6.

Shells from shellfish exploitation were also recovered, especially in trenches two and three. Trench 10 was found to be of a recent occupation with most of its pottery falling into the late Swahili and Post-Swahili periods that lie outside the scope of this work. Fractured stones were also recovered at the lower levels of trenches 3, 4, 5 and 9 (Fig.5.20, 5.22). Numbers of decorated pottery also decreased down the sequences in most of the trenches. Mnangole excavation provided the highest amount of crucibles, some of them with clearly cuprous staining inside (Fig. 5.25). Bases of some of them had been fired intensively to a stage of partial melting, possibly for smelting soft metal, although there was no recovered object related to the process.

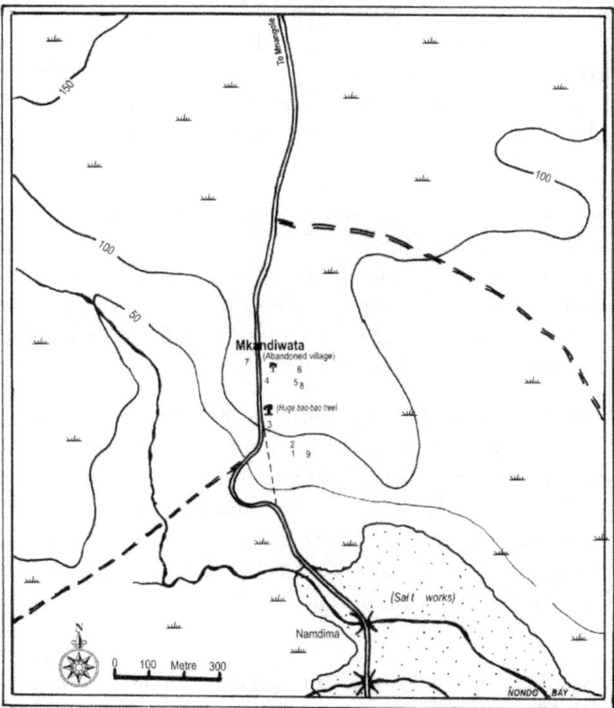

Fig. 5.21: Distribution of trenches (numbered 1-9) at Mkandiwata in Mnang'ole-Mchinga cluster.

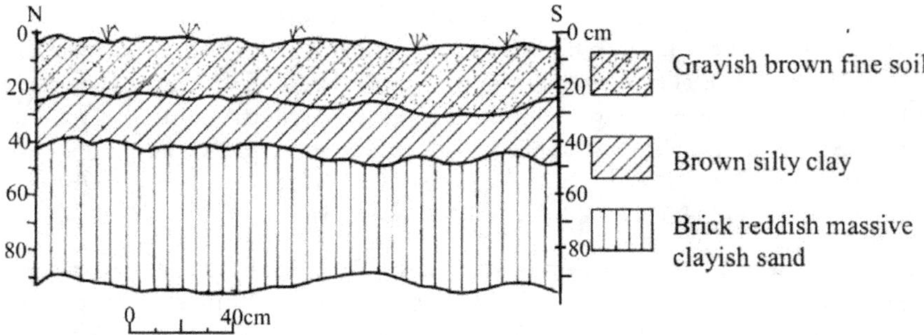

Fig.5.22: Eastern profile of trench 9 at Mnangole

Mkonya-ngonya Hill

Rushungi sites (39° 36.5'E, 9° 27'S) are located high up the hill 90m above sea level and about 1.5 km away from the nearest seawater. The underlying rock is nummulitic limestone, definitely from uplifted marine beds. This rock contains a very high percentage of calcium carbonate, mainly in the form of whole and fragmented circular-shaped shells of foraminiferid fossil called nummulites, cemented together with calcite. This rock is normally formed under marine conditions and in localised areas. The uplifting would have been far back in time during Pleistocene or even earlier.

In this research, an area of 34 square metres was excavated in four trenches in different locations of the settlement (Fig.5.23, 5.24). Almost 1200 pieces of pottery were recovered, of which 12% have decorations. An assortment of 39 beads was collected, including two unfinished blanks; most of them blue and Indian-red glass wound beads. Most of the recovered shells together with those forming the mound are of Telebralia palustric type of gastropods.

Trench 1, 3x2m in size, was opened five metres from the main road. The first level consisted of dark brown clayish soil that contained a few potsherds. The soil was yellowish brown, coarse and calcareous down the levels. As to the surface, shells were common throughout the sequence with maximum frequency in level 4. Two beads were recovered in level 2 (10-20cm). There was no significant change in term of pottery or shell species. The attributes of the pottery suggest a Proto-Swahili tradition (Fig.5.47-5.50). The sequence ended in level 6 (60cm), above the limestone basement rock.

Trench 2 of 2x3m and trench 3 of 2.5x4m, which are in line with the road, did not provide any significant change in terms of results. Trench 3, which is down the hill to the north of the first trench, was shallow (30cm) and had relatively less pottery.

84

Among the recovered 37 pottery pieces in this trench, only 4 had decorations. Trench 4 of 3x4m in size was richer in pottery and beads. About 134 decorated potsherds were recovered out of 860 pieces. 49 beads were also collected in association with this pottery that appeared to be part of the same tradition (Table 5.03; Fig. 5.31). Distribution of these beads is from the first level up to the fifth level, with the maximum in level 3 (n=32). The soil ends at level 7 (70cm) with residual nummulitic limestone.

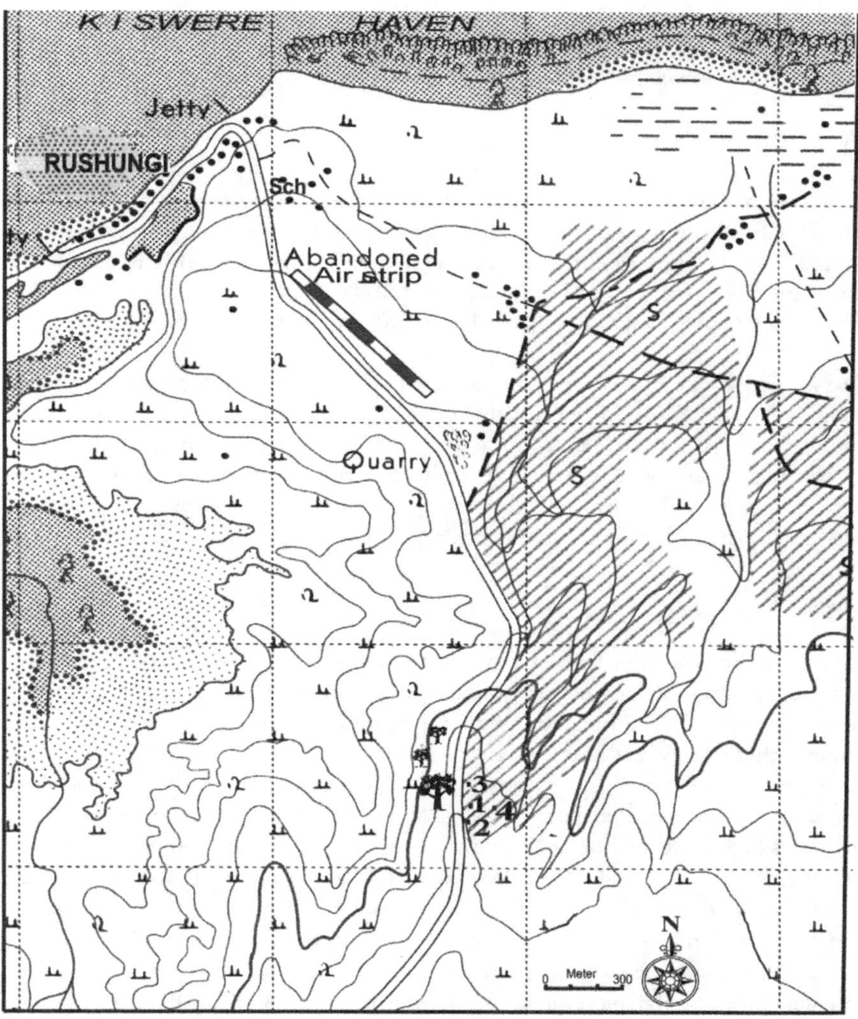

Fig.5.23: Locations of the trenches on Mkonyangonya hill along the main road to Mtandi

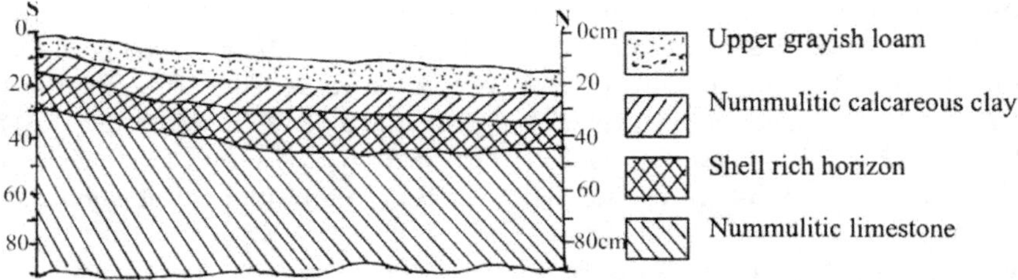

Fig. 5.24: Western profile of trench 1 at Rushungi

Kilwa Island

Excavations on the island were directed towards the areas that had shown surface distributions of pre-Swahili ware and scattering of fractured stone flakes or tools. Masakasa and Nguruni localities were discussed in this work and elsewhere, as potential areas for Stone Age remains (in Chittick 1974: 254). The areas seemed unoccupied by the stone-building civilization of the 13th to 16th centuries of Kilwa.

An area of 23.5 m² and 19m² was excavated at Masakasa and Nguruni respectively (Fig. 5.07; 5.25; 5.26). While distribution of cultural materials at Nguruni is only up to 60cm, pottery at Masakasa was found up to the 10th arbitrary level (100cm). Frequency of decorated pottery is generally low, with only 4% at Masakasa, but only 1.5% at Nguruni. The lowest level potteries are in rather smaller pieces and without decorated parts, which in most trenches appear below the EIW pottery.

Nguruni

The excavations at Nguruni were not as successful as expected. Some of the trenches were either disturbed, or they provided non-diagnostic materials. In trench 1, only one sherd from level 2 had a decoration in the form of graphite on the inner side, and another from level 4 that looked like a Proto-Swahili ware. Several pieces of iron slag occurred in the 3rd and 4th levels. Two rim-potsherds from the lower 4th level were bevelled, a characteristic of EIW tradition (Fig.5.51). Potteries from the 5th and 6th levels had no decoration at all, and were of relatively small sized thin vessels and weathered. Trench 2 had a Swahili ware in the first two levels but the rest of the levels had no diagnostic pieces. Trench 3 had pottery up to the 5th level with only a single piece from the 4th level that had a Swahili type of decoration. Trench 4 was not entirely diagnostic although there were pieces of slag at the 3rd and 4th levels, and possibly Proto-Swahili pieces in level 3. In trench 5, there was mostly Swahili ware up to the 5th level.

86

Masakasa

The first trench was 2x3m in size and with greyish fine silty soil up to the 3rd level where it became hard silt-clay. From the fifth level it became very hard clay down to the 10th level. Pottery occurred from levels 1-4, but none of it was diagnostic. Lithics were found from level 3 to the 8th. In trench 2, the soil was greyish fine sands up the 13th level. A Swahili ware dominated up to the 3rd level. Pieces of sgraffito ware were also collected between level 2 and 3 (Fig. 5.32, A-B). From level 4 there was no decorated piece except one piece with a red-wash and another red-slip. The shapes were possibly part of the Proto-Swahili Ware. From levels 5-7 the pottery had a TIW features (Fig.5.52). Lithic fragments existed from the first level down to the 13th level (130cm) depth where sterile whitish calcareous soil was encountered (Fig.5.54-5.55).

Trench 3 of 3x1.5m and its extension of 1x2m provided the best stratigraphy at Masakasa (Fig.5.25). The first three levels were composed of some Swahili ware with short necks and a narrow band of oblique lines or punctuates around the shoulder. The 4th and 5th levels were clearly of Proto-Swahili tradition (Fig.5.53). A big number of crucible fragments were collected, especially from level 5 (n=110), some with clearly a copper coat inside. In level 6, out of 50 potsherds, none had decorations. A few crucibles were also collected from this level. Level 7 provided a TIW pottery. Level 8 was of typical EIW, most likely Kwale tradition, based on false relief decoration. Levels 9-11 provided a few small fragments of pottery without decorations. They were relatively thinned walls as opposed to the EIW pottery above them, which were mostly thick.

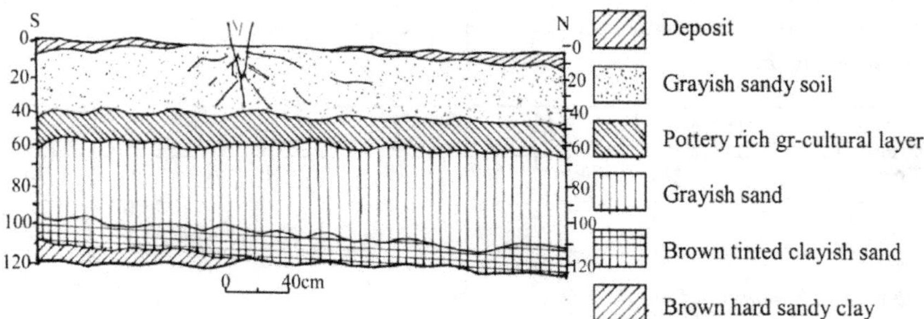

Fig. 5.25: Western profile of trench 3 at Masakasa 3

In the extension section, levels 3-5 provided additional Proto-Swahili ware. Level 6 had very fragmentary pieces (n=16) of weathered pottery and without decoration. A fragment of used crucible was also found in this level, possibly a Plain Ware horizon. Level 7 contained larger and thicker body pottery (n=30) of a TIW and some EIW bevelling. From level 8 was, again, a potsherd with a false-relief pattern of an EIW pottery. Below this level was only Lithics. Another trench of 2x1.5m was established close to the previous one. Levels 4 and 5 provided potsherds of Proto-Swahili ware; in level 6 a nicely bevelled thickened rim and some TIW potsherds, and level 7 had clearly pottery of EIW tradition.

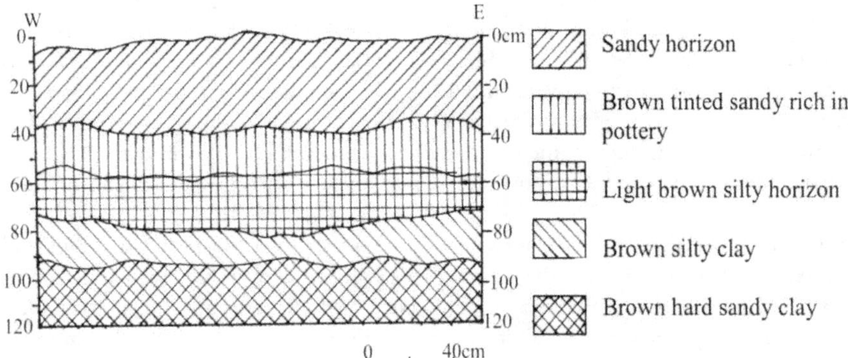

Fig. 5.26: Northern profile of trench 4 at Masakasa

Findings

The materials collected from each surveyed site and excavated trench are found in Tables 5.01-5.03. Pottery and lithic findings form the majority of the artefacts. More details concerning the distribution of different artefacts, designs and other pottery attributes in different trenches and levels are provided in the Appendices 1-III in the last pages of this book.

Grand	Trenches	Excav. Area	Depth	Pottery Total	// Decorated	Shells	Beads	Bones	Lithics
-Total/	53	235.1m²	130cm	19601	1928	2071	82	253	2015
			Max.	pcs.	pcs			frags	pcs
%				82%	10%	9%	0.3%	1%	8%

Table 5.01: Summary of the Research findings

Crucibles

These are made of grey, very friable clay, presumably refractory fabric, different from other pottery (Fig.5.27). The form of these crucibles in all cases is conical. The exterior and rim is usually vitrified to suggest intensive use at high temperature. There are occasionally traces of green cuprous layer at the inner bottom. Nearly all crucibles

88

are around 5cm in diameter measured externally. They are almost of the same kind in all the sites where they appear, especially Kilwa, Rushungi, Mnang'ole, but rarely in Mtwara sites.

Bones

The amount of bone fragments collected during the excavations on the Southern coast was very small, amounting to only about 1% of the total collection (Fig. 5.29). Most of this category came from the PSW horizon at Chikayowa site, and a few more from a similar horizon at Mnangole. The bones are fractured, suggesting food remains from the consumption of animal protein. The examination of the diagnostic pieces carried out by Prof. Terry Harrison identified only wild species including large fish (Fig. 5.30: A-C,), dik-dik and a small cat species. However, it cannot be denied that during the PSW tradition of the second millennium AD, people possessed domestic animals, although its evidence in the archaeological record of the Southern coast is still rare.

With this association of wild species, fish and shellfish, it became apparent that local people depended mainly on fishing and hunting for animal protein, probably to supplement their staple diet. There is not enough evidence to suggest that these people had acquired enough domestic livestock throughout the period under study although it is not surprising to find some.

Beads

The types and amount of beads collected are clearly summarized in Table 5.03, but also shown in Figure 5.31. The earliest of these beads come from PSW horizons, which have been thoroughly and absolutely dated. They mainly comprise Indian and some European glass beads, probably some of them produced locally at Rushungi. Two unfinished blanks (Fig.5.31: J, centre) are an indication that some of the beads could have been produced locally. Two of the red beads from Mnaida 3 (Fig.5.31: H) have a black core, possibly a European version. Most of the beads are from Rushungi and Chikayowa, which are typical PSW sites.

Shells

As discussed in section 5.1.4; marine shells, definitely from the exploitation of shellfish were found at most of the sites. At Mnaida 2, they were found even within the PIW horizons. Kabisela site showed a repeated period of intensive exploitation of marine resources (Fig.5.13), probably suggesting harsh periods such as drought and hence a shortage of other food sources. These shells also suggest a different behaviour from what is happening in the recent time, that people used to carry and process the shellfish at home rather than processing them on the shore. At Rushungi, for instance, shell heaps were found about 1.5km away from the shore, high up the hill. Some other sites showed similar patterns. Different species seemed to have been exploited depending on availability rather than desire, including Ostreidae (Oysters), Arcidae (Bivalves), Mytilidae (Mussels), Telebralia (gastropods) and varieties of crabs. A detailed work on shellfish exploitation on this Southern coast of Tanzania is best discussed in Msemwa's (1994) thesis.

Pottery and lithic artefacts

These two categories of findings are discussed in more details in the following chapters because they best provide clues and answers to the research questions of this work.

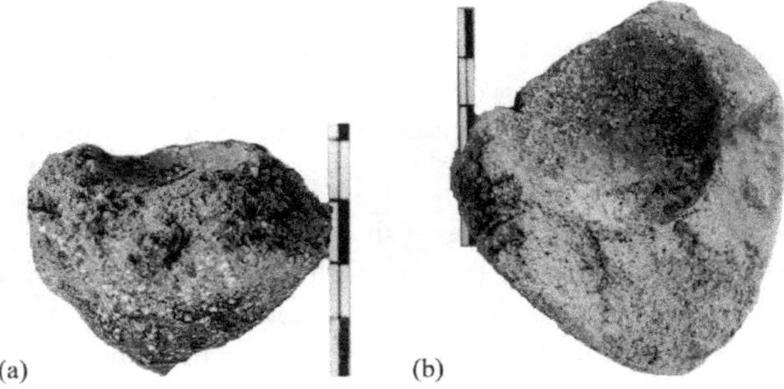

Fig. 5.27: Intensively used crucible bases with cuprous layer at the bottom from Proto-Swahili horizon at (a) Mkonyangonya Hill, Rushungi; (b) Masakasa, Kilwa Island.

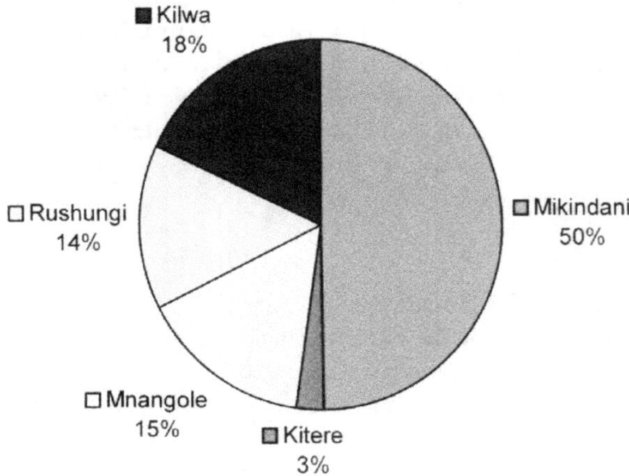

Fig.5.28: Sampling Intensity in terms of excavated area of each cluster during 2006 – 2007 (the total area is 235.1m²)

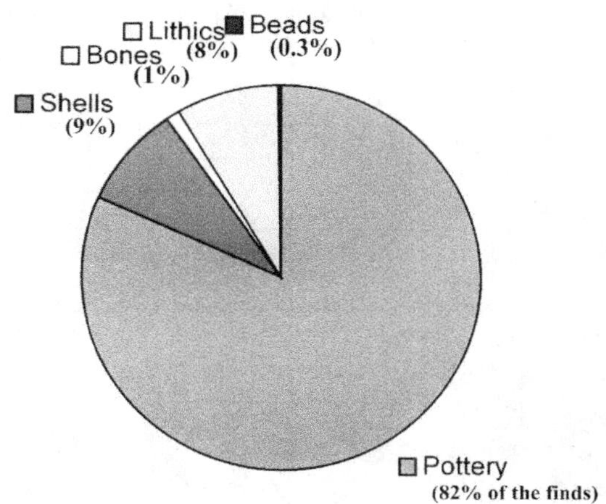

Fig.5.29: Percentage of different artefact types on the total collection from Southern coast during 2006-2007 field seasons.

Site name	Total	Decorated	%	R	R-S	S	R-N	RNS	N-S	Body sherd	%
Mikindani cluster:											
Mnaida 1	28	2	7.14	0	1	2	0	0	0	25	89.29
Mnaida 2	831	77	7.27	14	23	24	4	8	5	753	90.61
Mnaida 3	1305	146	11.19	17	20	41	0	2	5	1220	93.49
Nakatumbatu	828	3	0.36	6	18	1	2	9	1	791	95.53
Kabisela	305	72	23.61	26	26	44	0	7	7	195	63.93
Chikayowa	3701	397	10.73	166	111	186	16	10	47	3165	85.52
Pemba	324	53	16.36	10	15	18	5	15	4	257	79.32
Kitere	157	29	18.47	9	8	13	6	0	4	117	74.52
Mnang'ole	6077	649	10.68	232	82	454	24	12	27	5246	86.33
Rushungi	1234	159	12.88	50	34	61	1	13	24	1051	85.17
Masakasa	2250	88	3.91	29	45	59	8	24	34	2051	91.16
Nguruni	1134	17	1.50	11	19	28	2	15	17	1042	91.89
Total	19601	1928	9.84	570	402	931	68	115	175		

Table 5.02: Pottery Distribution on different sites

Abbr: R – rim pieces; S– houlder; N-neck; R-S –rim-shoulder; R-N - rim-neck; RNS- rim-neck-shoulder; N-S- neck-shoulder (pieces).

	Mnaida	K'sela	C/yowa	Nakat	P'mba	M'ngole	Rushun	Masak	Ngur	Total
Red	2	2	12	0	0	1	15	1	0	33
Blue	1	0	4			1	30	1		37
Green		0	1			1	5		1	8
Yellow		0	0				1			1
White	2	0	1							3
Total	5	2	18	0	0	3	51	2	1	82

Table 5.03: Beads from the year 2006-2007 Southern coast excavations

Abbr: (second row) Mnaida, Kabisela, Chikayowa, Nakatumbatu, Pemba, Mnang'ole, Rushungi, Masakasa and Nguruni

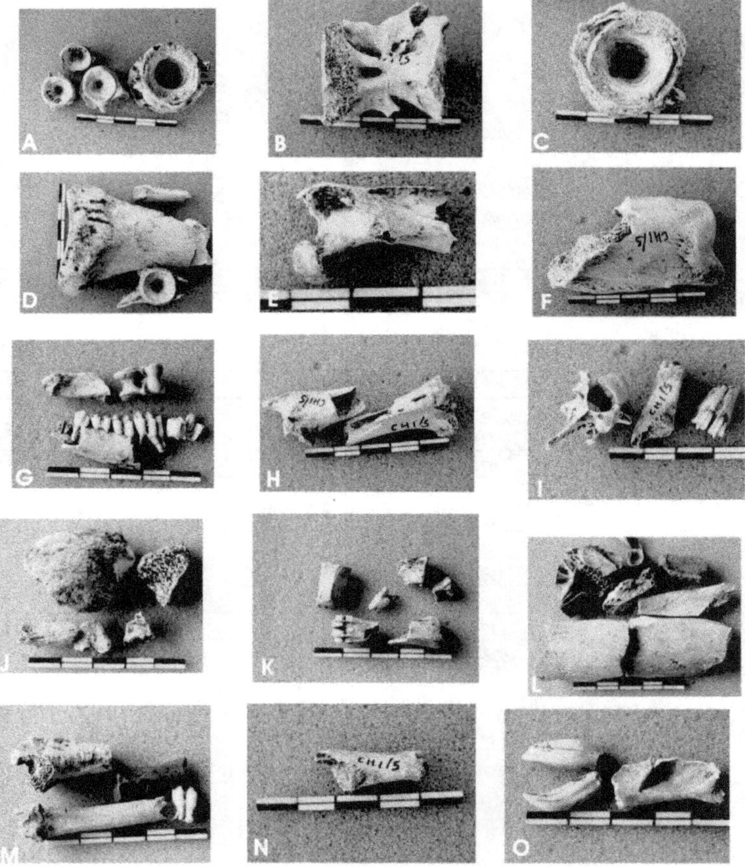

Fig. 5.30: Views of food-bone remains from Chikayowa: A-C, D (bottom), L (top-left) are fish bones. G-I, K, N, are certainly wild dik-dik. J (bottom-left) and M (upper-right) are dik-dik horn cores. O is carnivore

92

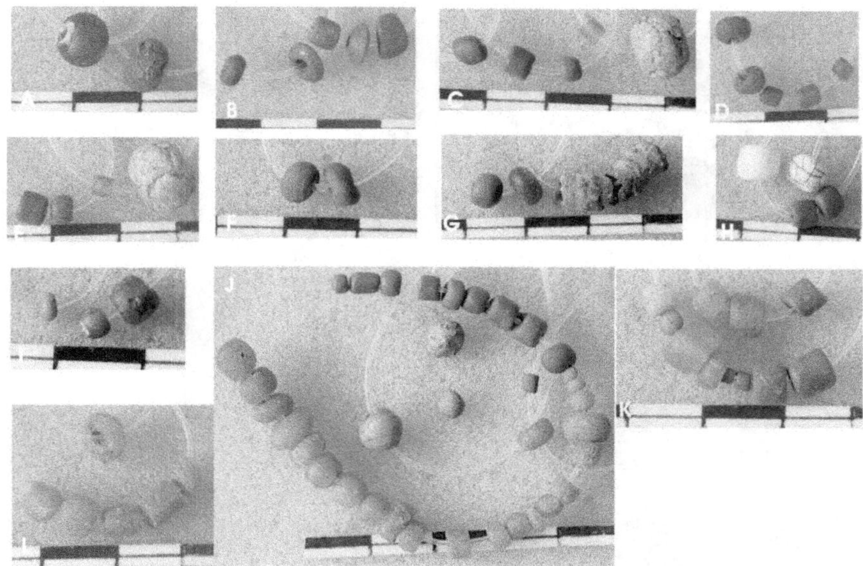

Fig.5.31: Views of beads Assortment from Southern coast: A,=L2, B,D,=L3; C,E,=L4 (Chikayowa); F,=Kabisela L4, G,=Kilwa, H,=Mnaida3, I,=Mnang'ole, J,=Rushungi L3, K,=Rushungi L3&L5, L, =Rushungi L2.

Fig.5.32: Views of some imported and local ceramics from Kilwa (A-B) and Kabisela (C).

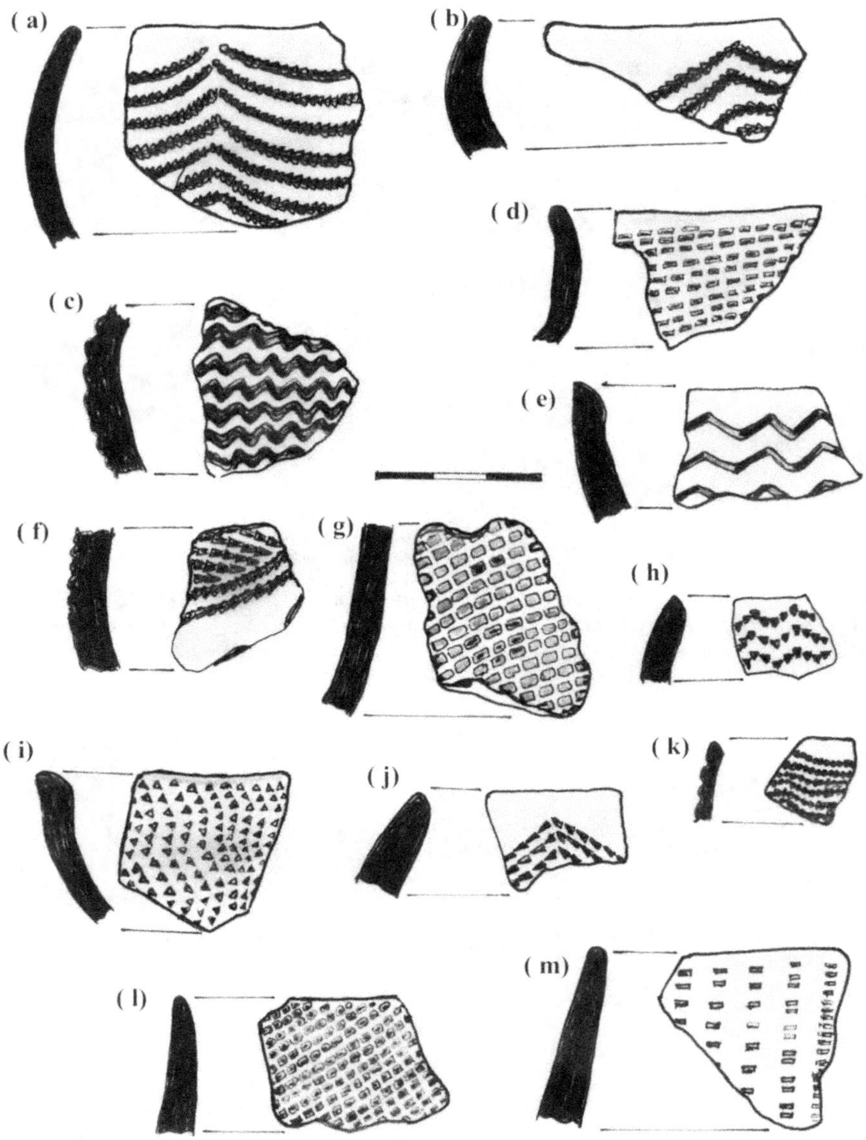

Fig.5.33: Mnaida PIW pottery

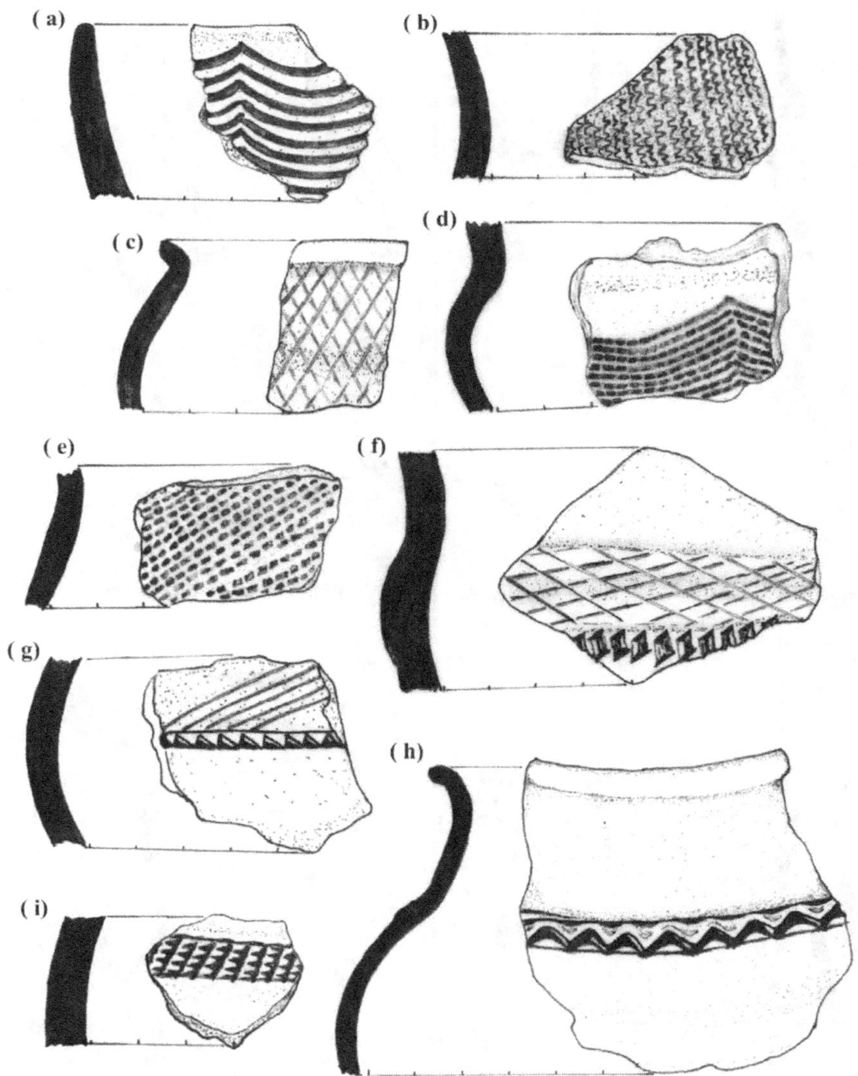

Fig. 5.34: PIW pottery from Mnaida

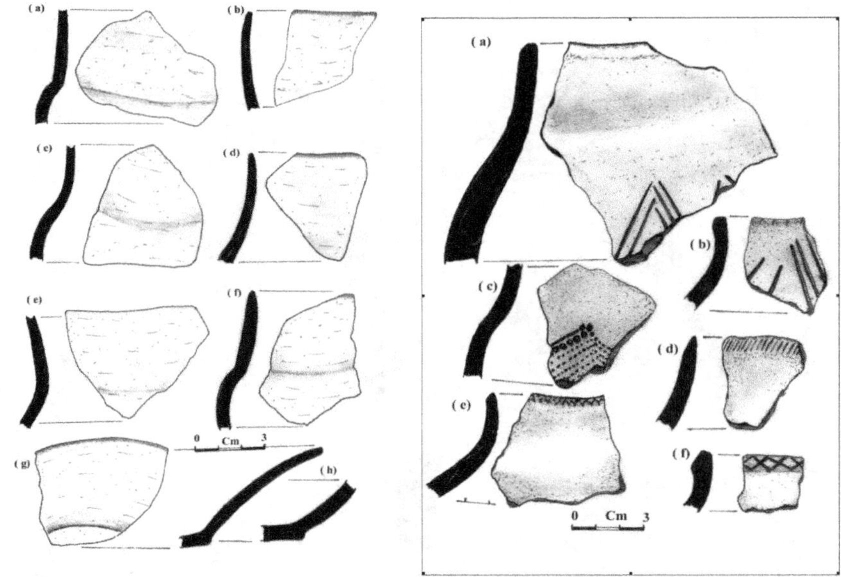

Fig.5.35: Plain ware from Nakatumbatu and Kabisela   Fig.5.36: TIW pottery from Kabisela

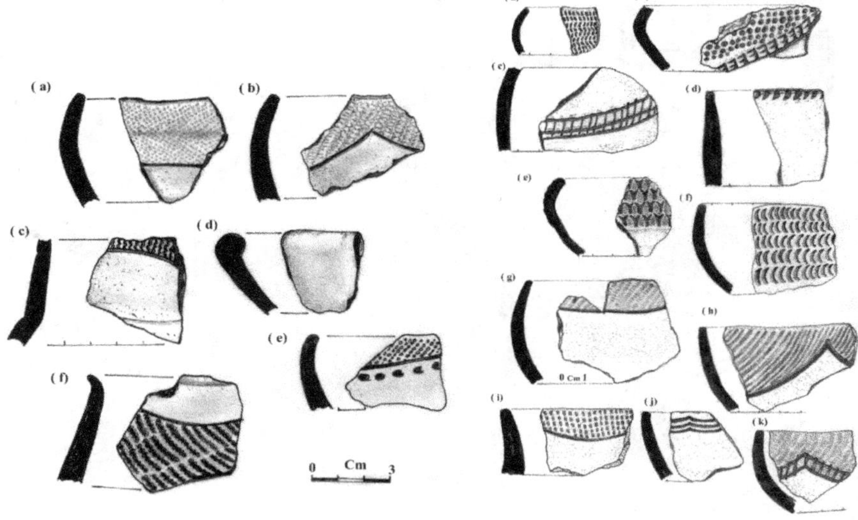

Fig.5.37: Proto-Swahili ware from Kabisela   Fig.5.38: Chikayowa Proto-Swahili ware

96

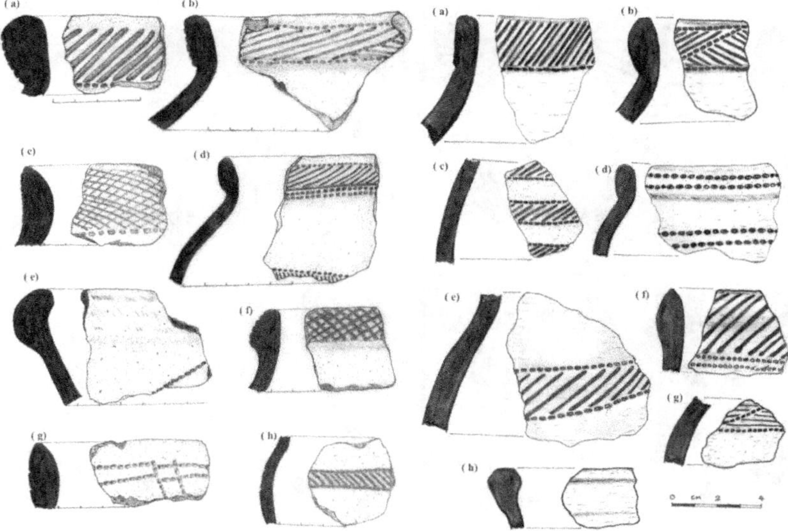

Fig.5.39: Mikindani EIW pottery

Fig.5.40: Mikindani EIW pottery

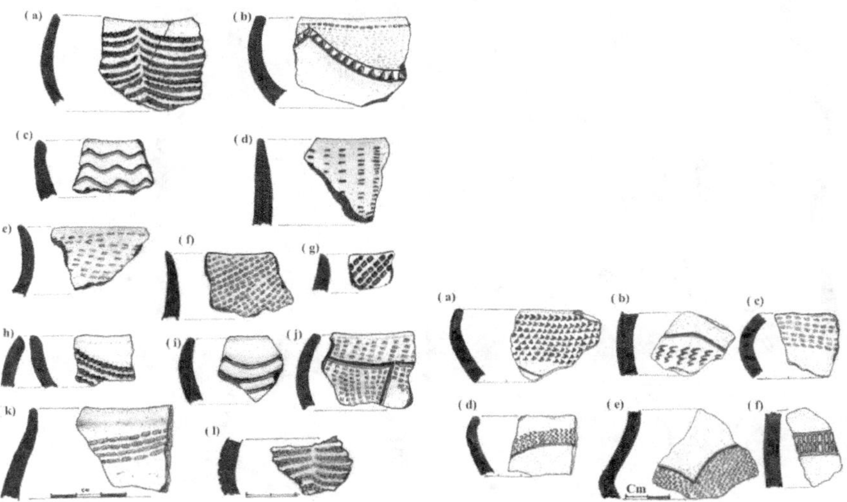

Fig.5.41: Kitere PIW ware

Fig.5.43: Kitere Proto-Swahili ware

97

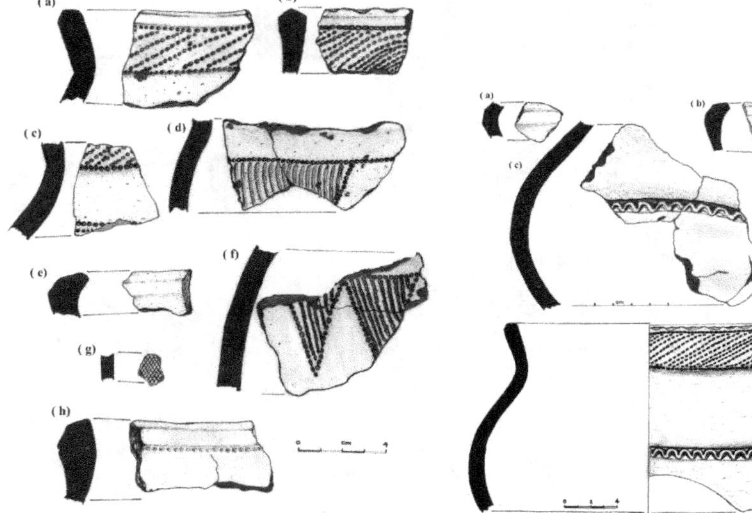

Fig.5.43: EIW pottery from Mnang'ole

Fig.5.44: EIW pottery from Mnang'ole

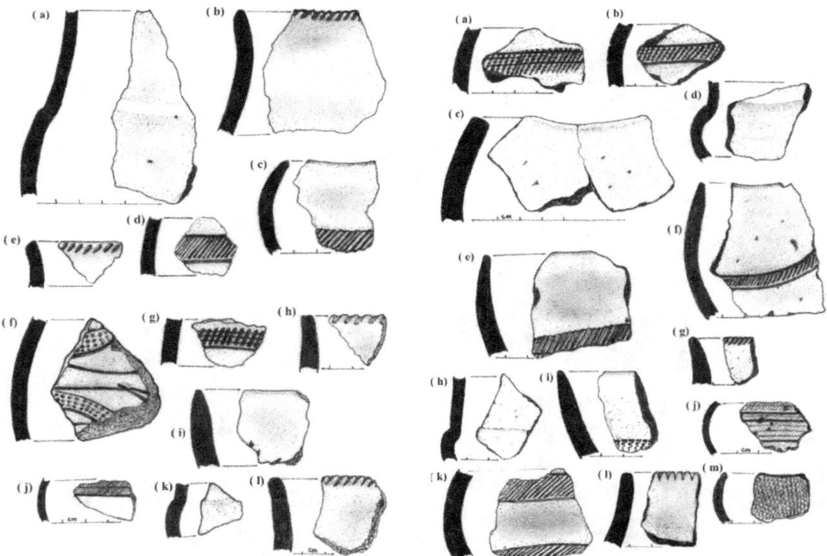

Fig.5.45: Proto Swahili ware from Mnang'ole

Fig.5.46: Proto-Swahili ware from Rushungi

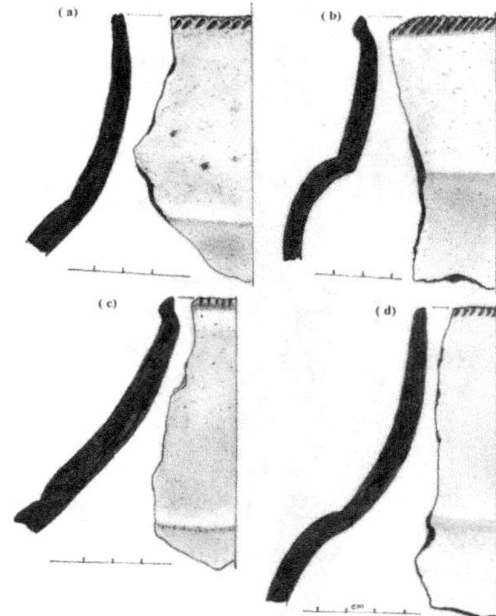

Fig.5.47: Some of the Proto Swahili necked forms from Rushingi

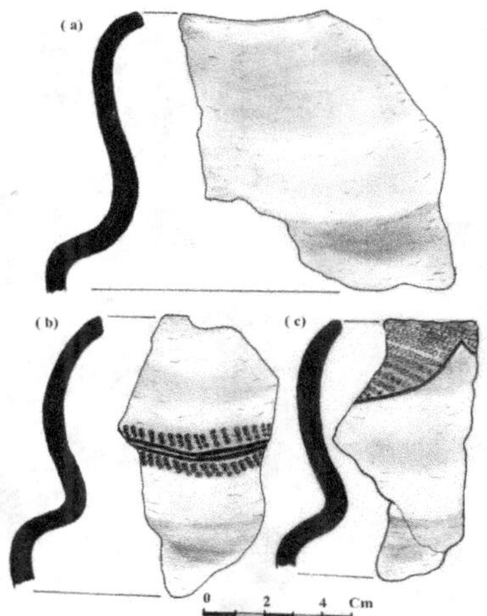

Fig.5.48: Double-curved forms in Proto-Swahili ware of Mnangole

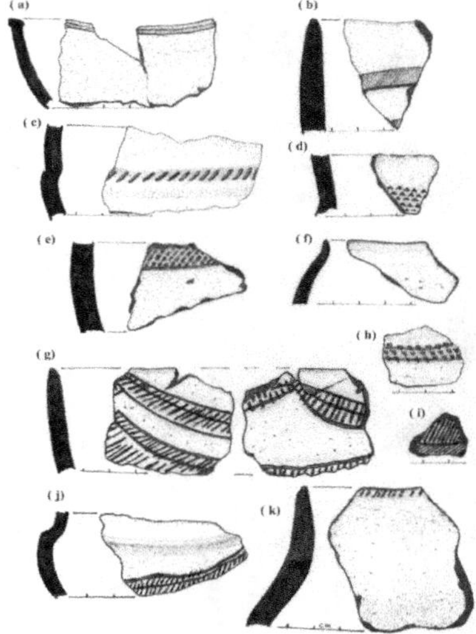

Fig.5.49: Proto-Swahili ware from Rushungi

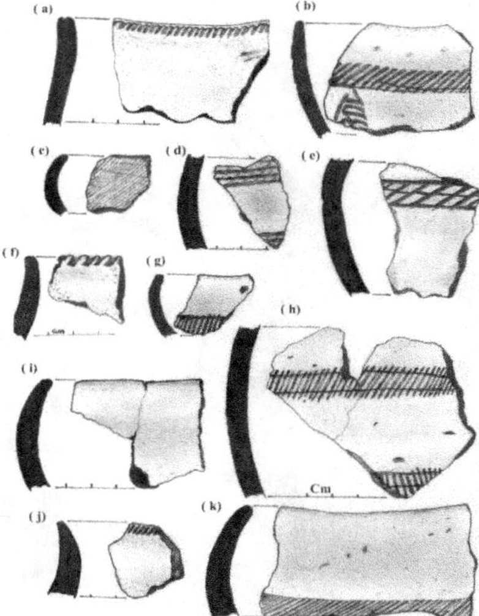

Fig.5.50: Proto-Swahili ware from Rushungi

99

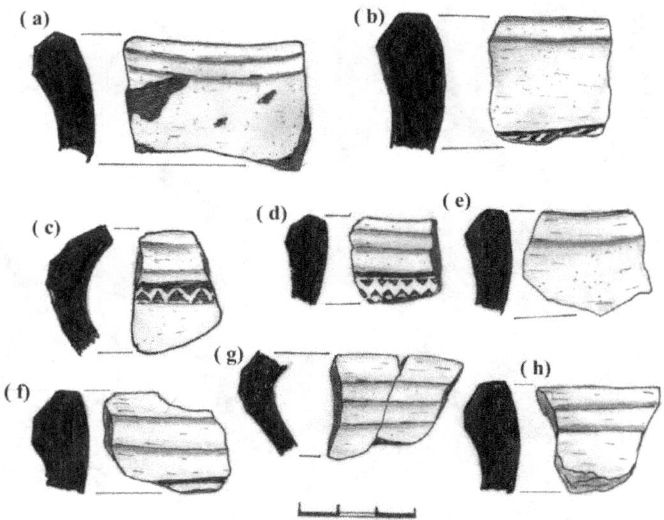

Fig.5.51: EIW pottery from Masakasa, Kilwa Island

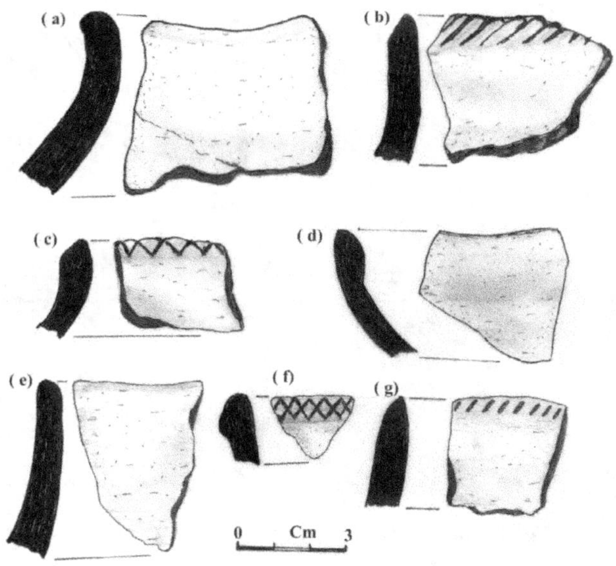

Fig.5.52: TIW pottery from Masakasa, Kilwa Island

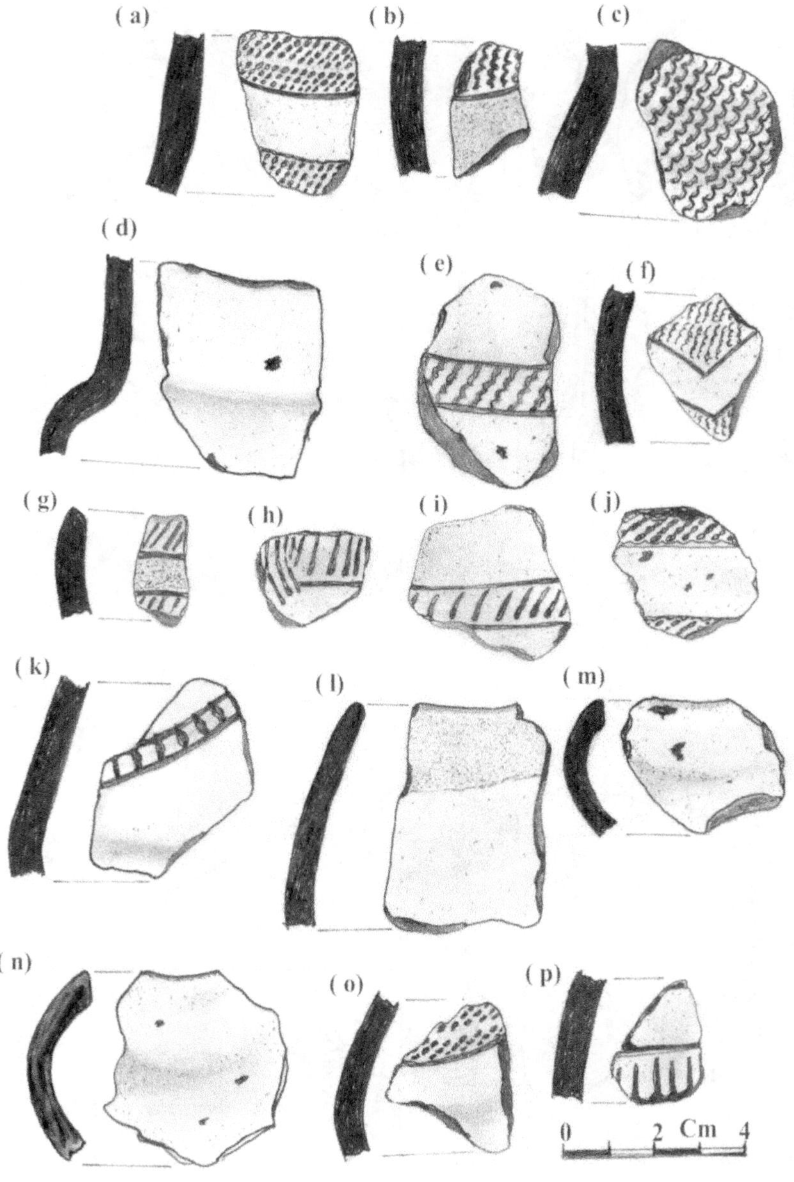

Fig.5.53: Proto-Swahili ware from Kilwa Island

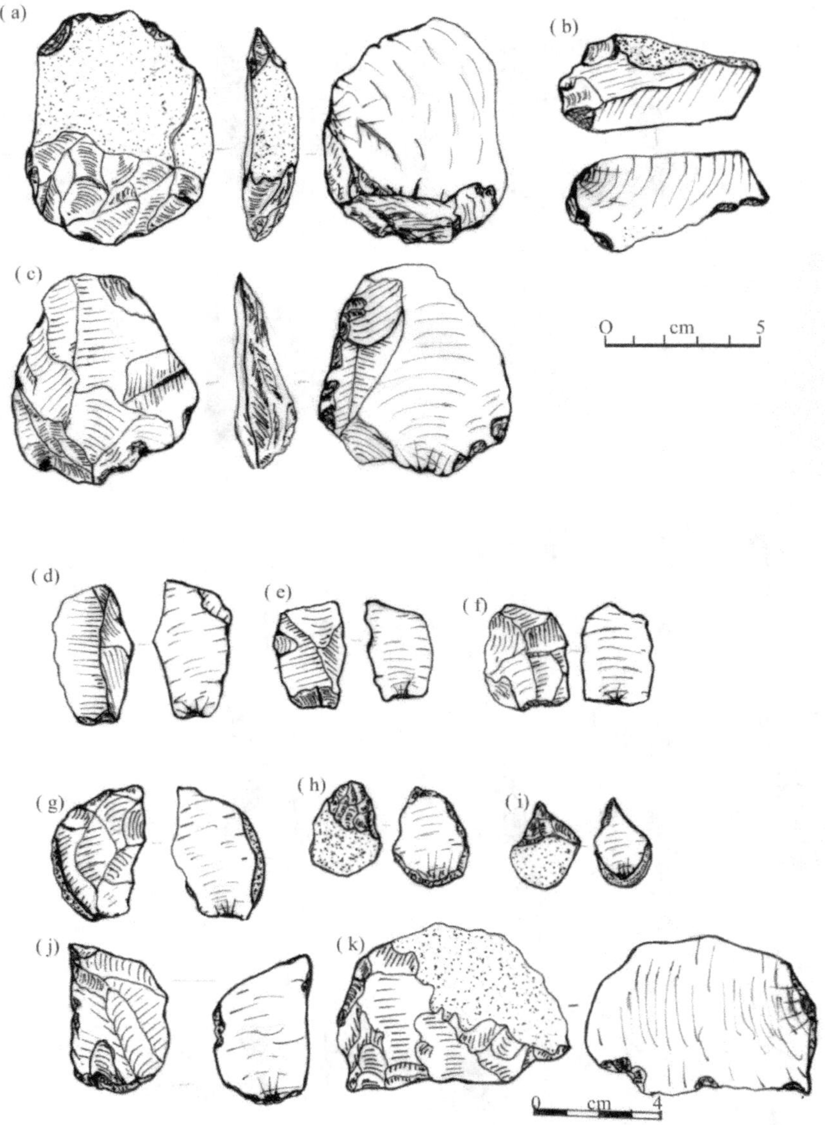

Fig.5.54: Stone artefacts from Kilwa including convex end scraper (a), blades (b, d-f), unifacial point (c), curve backed piece (g,) small bipolar core-fragments (h-i), Convex side-scraper (j-k).

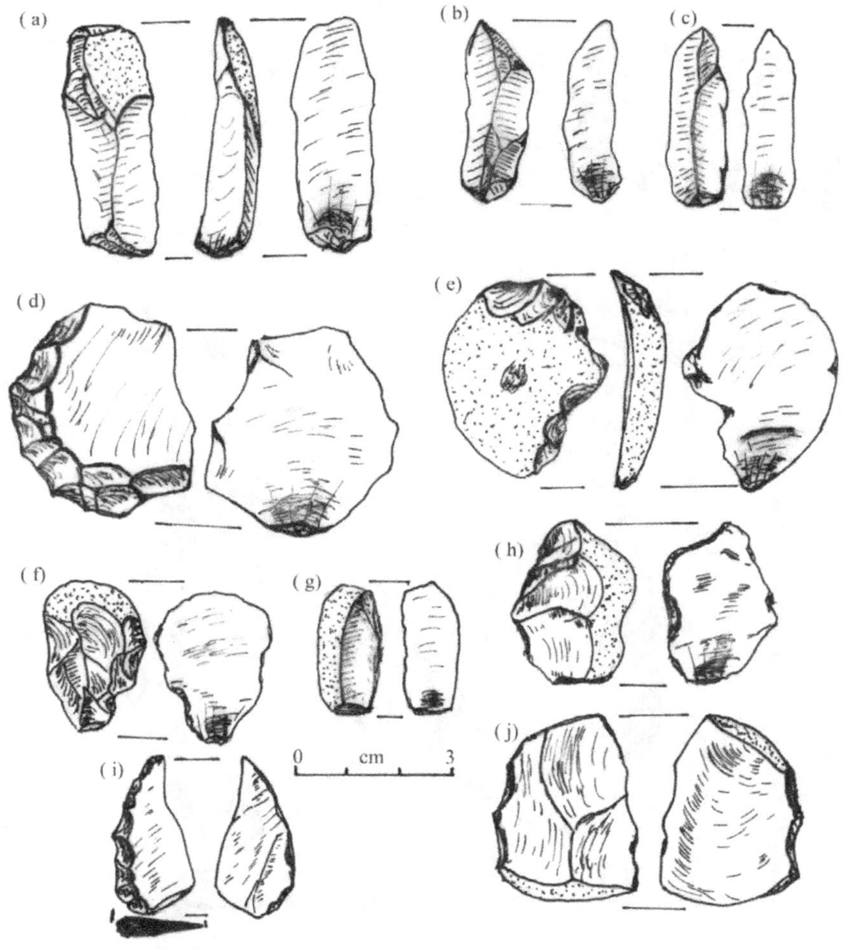

*Fig.5.55: Stone artefacts from Kilwa including blades (a-c, g); curve blacked piece (d,i)
utilized/trimmed flake (f, h, j); circular scraper (e).*

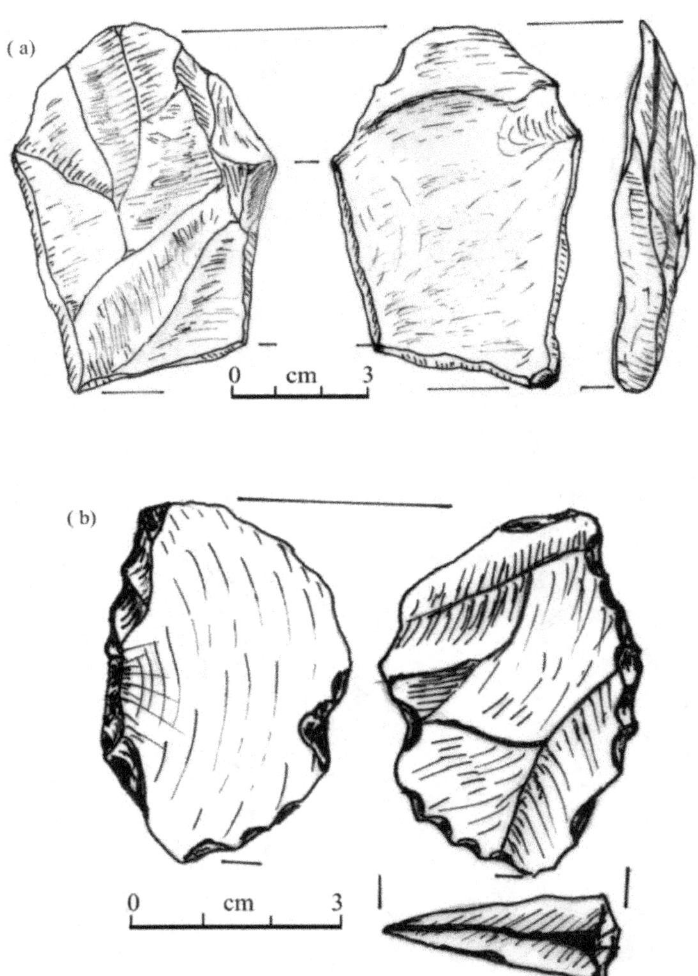

Fig.5.56: Stone tools from Mnangole (a) and Mnaida 3 (b): convex end scraper (a), curve-blacked piece (b)

104

Chapter **6**

POTTERY AND LITHICS ANALYSES

Introduction

It has been shown in Chapter Five that pottery is by far the most frequent artefacts found in early settled sites. This answered a curious question from one of the archaeology students on why this work was so interested in potsherds. Reasons behind this phenomenon are more obvious, that ceramics are far more resistant and durable in different conditions of soils than other synthetic materials. As argued at the beginning of this work, the major concern is on evidences that support early coastal occupation by man.

The presence of shell accumulations has been regarded as refuse from the exploitation and consumption of marine resources. Their presence has been used as a good indicator of a site of human activity (Msemwa 1994), although not as a direct evidence for a settled life. However, species variations in the shell mound sites have been interpreted as a function of the environment and the availability of species rather than of human behaviour. Hence, excavations of these types of sites have tended to recover associated artefacts such as pottery, lithic or other man-made objects than the shells themselves (in Msemwa 1994). Based on the above observations, this work focuses on pottery to establish different settled communities in time and space, and a little more on lithic assemblages, albeit limited in their occurrence.

Many scholars have discussed the question as to whether pottery can be used to identify a tradition or a cultural or ethnic group (Soper 1971, Huffman 1980, 1989, Collett & Robertshaw 1983, Rice 1987). Huffman (1980), for instance, has argued that pottery reflects human craft that carries a group rather than an individual identity. What is essential in this chapter is the fact that throughout the last millennium BC and first millennium AD, communities which were producing pottery that bears a broader spatial relationship with time, occupied the southern coast of Tanzania and its hinterland. Over a period of time, the technology changed or evolved, with new craft elements replacing the old in whatever mechanism under internal or even external influences.

A few aspects should be noted at this stage: Whether these communities belonged to the same ethnic group or not, the important aspect is that they had some cultural similarities or economic contact. One of the principal aims of this work has been to study the decorations and forms of these potteries in order to obtain information on their users' origins, and the spread and degree of cultural and economic affiliations.

Secondly, some arguments relating to this work have been reported concerning the PIW sites of the southern coast of Tanzania (Chami & Kwekason 2003, Chami 2006: 100, Kwekason 2007). The early pottery found on this coast has been analysed and compared with counterparts from the Rift Valley of eastern Africa. The pottery was tentatively found comparable to the Nderit tradition of the Rift Valley that dates from 7000 – 3000BP (Bower et al., 1977, Robertshaw 1990) and Narosura of about 3000 – 2300BP (Odner 1972, Bower & Nelson 1978). Wandiba and Kusimba have recently viewed this pottery as related to the Kenyan Rift valley Late Stone Age pottery from what Kusimba is calling "Stone Age settlements" (pers. comm., Nov. 2009 in Mombasa).

Thirdly, the main decorative motifs of the EIW tradition have been discussed in detail elsewhere (Fagan 1961, Soper 1971, Huffman 1970, 2005, Phillipson 1975, 1976, Chami 1998). For the TIW, Plain Ware and Swahili ware, the best reference is made to Chami's (1994, 1998). Proto-Swahili tradition (PSW) finds its only reference in Chittick's work at Kilwa (1966, 1974) where it forms part of the so-called "Kitchen Ware". Re-examination of the pottery and its comparison with the newly established archaeological sequence suggests three distinct traditions in the "Kitchen Ware". The lower material is of TIW tradition, largely corresponding to the "Period 1a". The middle level comprises "Plain Ware" tradition that is represented by a few, but distinctive pieces of pottery. The upper part of the collection has been attributed to the PSW tradition, corresponding partly to the "Period 1b" and "Period II' in Chittick's (1966, 1974) work. Its age ranges from the 11th up to the early 14th century AD.

Analytical Attributes of Pottery

It has been argued that success of an archaeological enterprise depends on our learning to ask right questions at the right moment, and finding the most productive means of answering them (Renfrew & Bahn 1996). The analytical attributes of archaeological artefacts vary, and depend on the object and objectives of the analysis. Common analytical methods are better discussed in more sophisticated literature like that of Shepard (1971), Rice (1987) and others. Different archaeologists have, however, used different attributes in their studies, depending on the nature of their research problems.

Soper (1971) used an eclectic approach by selecting a wide range of attributes of shape, rim morphology, base-shape, surface finish, decoration techniques and motifs to clarify the EIW pottery of East Africa. Huffman (1976, 1980) concentrated on the structure of motifs and decoration placements to classify the pottery of southern Africa. Collett and Robertshaw (1983) used decorative motifs and shape to differentiate the PIW pottery of the Rift Valley and other sites of East Africa. In this exercise they found decorative techniques less informative and less distinctive in providing clues about a tradition (ibid. p.109). Chami (1994) combined decorations, shape, motif placements and fabric to define the TIW of eastern Africa. Decoration, shape and rim-morphology were also most useful attributes in clarifying the Swahili archaeology of the coast of East Africa (Chami 1998).

In this work, decorations, shape, rim morphology, motif placements, and decorative techniques are combined to define different assemblages from the Southern coast. The aim is to find out how these attributes could be distinctively used to differentiate one assemblage from another with respect to time and space. It also paid attention on the argument that specific identity of the designs – the content, is not so important but the relationship between designs. Meaning arises not from any specific design in isolation but from the relationship between different designs (Tilley 1991:105).

Decorations

Until now, there is no standardised way of defining decorative structures that appear on pottery. Different scholars have subjectively used different terminologies to describe these structures depending on their personal experience and background. The term "motif" appears to be a more popular word used to refer to these structures (Huffman 1976, 1980; Collett & Robertshaw 1983; Soper 1971), although there are other terms like "decorative format" (Sinclair 1987). Shepard (1971) has gone further by defining "element" as the smallest decorative structure that can either repeatedly and independently form a particular pattern, or can be combined to form a motif that makes a more visible pattern. All in all, the study of these individual elements and motifs has been less successful in pottery classification because they have been used differently in time and space. The focus is now on the associated forms and designs at a specific time (Tilley 1991). However, graphic presentation of the decorative structures is always considered primary, while verbal comments remain secondary (Shepard 1971). For that reason this work invested a lot in graphic illustrations.

A wide range of design patterns was identified; representing merely four cultural phases of, in this matter, traditions (Fig.6.01, Table 6.04). These include what is here defined as Pre-Ironworking ware (PIW), Early Ironworking ware (EIW), Triangular Incised ware (TIW) and Proto-Swahili ware (PSW), ignoring the Swahili ware which is also represented. A total of 76 more or less different design patterns were identified and their spatial and temporal distributions established. The numbers and percentages of decorated sherds from all excavated sites, ranging from PIW to PSW traditions are shown in Table 6.01 and its comparisons in Figure 6.04 and 6.05.

The Kabisela site was found to have the greatest percentage of decorated sherds although the sample size was relatively small, followed by Kitere and Pemba sites (Fig.6.05a). Nguruni site on the other hand, though with a high concentration of pottery, showed the lowest percentage of decorated sherds, followed by Masakasa, both on Kilwa Island. However, the sites that provided PSW ware had a large number of both undecorated and decorated pottery. More than 91 % of the Kilwa pottery comprised body sherds while Kabisela had the lowest number of body sherd pieces (Fig. 6.05a).

Rushungi site had a highest percentage of sherds burnished with graphite compared to other decorative techniques, followed by Nguruni and Mnangole (Table 6.03, 12th row) although when compared to other sites Mnangole provided larger amount (Table 6.03, 9th column). Indented rims were mostly confined to the vessels with constricted neck in PSW ware, where Rushungi and Mnangole have higher frequencies, followed by Chikayowa. The area of decorations has an insignificant range, although a few sherds had both external and internal decorations especially in PSW ware. Decoration techniques varied from impressed decorations in the form of stamping and punctuates to medium and coarse incised decorations. There was no sign of local glazing in all the sites and the proportion of glazed pieces was very small. Kilwa and Rushungi sites had a greater percentage of PSW with incised decorations than the sites further south that had more stamps (Table 6.03). The TIW decorations from Kilwa in the north and Kabisela in the south were more or less quite similar.

There was significant variation in the EIW pottery from Kilwa, Mnangole and Pemba sites. From Kilwa it rarely consisted of the incised or stamped band of decoration but instead a line of false relief chevrons was more common. At Mnangole the decoration band was usually oblique lines of comb-stamps, in one case in combination with a line of false relief chevrons. At Mikindani, the decoration band was made of incised oblique or horizontal lines. The shoulder triangular pendants at Mnangole were filled with vertical incised lines while at Mikindani they had horizontal lines. The corners in the bevelling became more perfect or sharper in the north than in the south, where the edges are more or less blunted (Fig.6.03). Mnaida tradition showed more rocker-stamping decorations than any other tradition.

The design motifs of most decorations are summarized in Table 6.05. The motifs cluster in the four groups representing PIW, EIW, TIW and PSW, as discussed below.

PIW: The characteristic of this ware is the decoration of the surface with mainly impressed cuneiform patterns. The most typical pattern is the festooned band of parallel denticulate lines executed by rocker stamping using a triangular pointed object. In some instances, closely packed dots or angular points make the lines in the similar pattern. In most cases the festoons are formed by a series of curved pieces of denticulate lines rather than a continuous loop (Fig.5.33; 5.41).

In other instances, the cuneiforms form lines or a horizontal band around the rim-shoulder section. Other characteristic decorations include regular packed stamps of square or rectangular shapes. A few specimens showed a chevron or wavy incised lines around the shoulder (Fig.5.33: c, e, h, i). One of the specimen displayed a line of false relief chevrons below the base of the neck (Fig. 5.34: h). Two others showed a combination of hatching and either a cuneiform or 'lozenge' line on the shoulder. Mnaida hill contains the best-preserved remains of this type of ware, which have been described as Mnaida tradition in this work.

Kitere on the hinterland is another site that clearly contains this tradition (Fig. 5.41). While the Mnaida tradition displays more features that resemble those of the Nderit ware of the Rift Valley, towards the north in Kilwa and Mbwemkuru valley there is slightly more emphasis on band decorations like those of Narosura of the Rift valley than on cuneiform and unbounded stamps. The excavated Kilwa materials were too fragmented and weathered to come up with a better comparison, hence the comparison made here involved surface collection.

EIW: In this group, the decoration is mainly incised oblique lines commonly bounded by a dotted line or punctuates. In some instances double lines of punctuates exist on the lower side. Occasionally the bounding line is missing on the upper part and rarely on both sides (Fig.5.39-40). At Mnangole site the decoration band is filled with impressed comb-stamps, set at an angle, rather than incised lines (Fig.5.43). Either of the bands is rare at Kilwa, where a line of false relief chevrons is more common just below the band of mainly three bevels. Other characteristic decorations include a series of triangular pendants hanging around the shoulder. These pendants are filled with horizontal lines especially at Mtwara sites, or vertical lines are common at the Mnangole site.

In some instances, a line of false relief chevrons replaces the pendants, as witnessed at Mnangole (Fig. 5.44, 6.08). On one of the specimens, an unusual combination of bevels and broad denticulate on the lip, stamps on the rim and false relief chevrons on the shoulder were displayed (Fig. 6.08; left). A few samples at Mnangole display features similar to those at Kilwa, with up to three narrow bevels sometimes combined with a line of false relief chevrons below the bevels. In a few instances, especially at Mikindani, incised band or double lines of punctuates is displayed around the shoulder instead of pendants. False relief chevrons are entirely missing in Mikindani sites.

TIW: Characteristic decorations are hatching on the outer side of the rim. This may consist either of strokes set at an angle or of cross-hatching. This kind of hatching is found on the rim itself of this vessel type only. Impressed lines of punctuates are also found at Kabisela site (Fig. 5.36). On the neck or usually just below the neck, the commonest motif is the triangle, the base being at approximately the height of the top of the shoulder of the pot. They may be hatched or incised, but in rare cases with impressed denticulate lines.

PSW: This group has a wide range of decorations (see Table 6.04) but mostly in the form of festooned, looped or even horizontal bands filled with incised, impressed or burnished decorations (Fig. 5.37-38). More common is the arched band that starts at the base of the neck towards the rim or the festoon hanging from the upper part of the rim. The horizontal bands are in most cases on the upper part or rim-shoulder region, or just around the shoulder. Cross-hatching is also common at Kitere site, usually extending up from the base to the top of the rim. Here, the grooves are very fine, narrow and shallow, executed in leather-hard condition.

At Kabisela, Chikayowa, and occasionally at Kilwa, the band of decorations, either in the form of horizontal lines or festoons extend from the upper part of the rim to the shoulder, while at Rushungi and Mnangole sites it is mainly in the form of hanging loops, chevrons or horizontal bands around the shoulder (Fig. 5.45-46, 5.49-50). Denticulate is the only prominent decoration on the upper part of the rim and only on necked forms. Undecorated parts of the pots in this group are sometimes burnished with bright graphite.

Site name	Total	Decorated	%	R	R-S	S	R-N	RNS	N-S	Body sherd	%
Mnaida 1	28	2	7.14	0	1	2	0	0	0	25	89.29
Mnaida 2	831	77	7.27	14	23	24	4	8	5	753	90.61
Mnaida 3	1305	146	11.19	17	20	41	0	2	5	1220	93.49
Nakatumbatu	828	3	0.36	6	18	1	2	9	1	791	95.53
Kabisela	305	72	23.61	26	26	44	0	7	7	195	63.93
Chikayowa	3701	397	10.73	166	111	186	16	10	47	3165	85.52
Pemba	324	53	16.36	10	15	18	5	15	4	257	79.32
Kitere	157	29	18.47	9	8	13	6	0	4	117	74.52
Mnang'ole	6077	649	10.68	232	82	454	24	12	27	5246	86.33
Rushungi	1234	159	12.88	50	34	61	1	13	24	1051	85.17
Masakasa	2250	88	3.91	29	45	59	8	24	34	2051	91.16
Nguruni	1134	17	1.50	11	19	28	2	15	17	1042	91.89
Total	19601	1928		570	402	931	68	115	175	15916	
%		9.84		2.91	2.05	4.75	0.35	0.59	0.89	81.2	

Table 6.01: Proportions of different sherd types on sites

Abbr: Rim(R); Rim-Shoulder (R-S); Shoulder (S), Rim-Neck (R-N), Rim-Neck-Shoulder (RNS), Neck-Shoulder (N-S)

Site name	IG	%	EG	%	ID	%	RD	%	SD	%	LD	%	Total
Mnaida 1	0		0		0		2	66.7	1	33.3	0		3
Mnaida 2	2	2.6	0		1	1.3	40	51.9	34	44.2	0		77
Mnaida 3	3	2.1	1	0.7	3	2.1	49	33.3	91	61.9	0		147
Nakatumbatu	0		0		0		0		3	100	0		3
Kabisela	1	1.4	3	4.1	0		9	12.3	58	79.5	2	2.7	73
Chikayowa	14	3.5	8	2	1	0.25	85	21.2	248	61.7	46	11.4	402
Pemba	1	1.8	1	1.8	0		28	51.9	24	44.4	0		54
Kitere	2	4.2	1	2.1	0		19	39.9	23	47.9	3	6.3	48
Mnang'ole	55	8.4	57	8.7	3	0.46	25	3.8	405	62.1	107	16.4	652
Rushungi	11	29	12	7.4	1	0.6	5	3.1	86	53.1	47	29	162
Masakasa	3	3.4	6	6.8	0		23	26.1	49	55.7	7	7.9	88
Nguruni	3	18.7	2	12.5	0		1	6.3	10	62.5	0		16
Total	95	5.5	91	5.3	9	0.5	284	16.5	1028	59.8	212	12.3	1719

Table 6.02: Variations in area of decorations

Abbr: IG= Interior graphite; EG= Exterior graphite; ID= Interior decorations; RD= rim-decorations; SD= Shoulder decorations; LD= lip-decorations

There are some observable inter-site variations within the tradition, especially in the placement and filling of the decoration bands. At Mnaida 3, for instance, the majority of the decorations are in the form of looped bands on the shoulder, filled with comb stamps. At Chikayowa, comb stamping is mainly as festoons or a wide band on the rim-shoulder area. Loops are mainly formed with parallel lines. Shell-impressed bands are also common.

At Kitere, the majority are wide bands of comb stamps and arched bands of impressed shells or combs on rims or shoulder. Shell-impressed decorations are more common at Mnangole site. They are displayed as festoons, arched or looped bands. Incised and stamped patterns also exist in reasonable amount. At Rushungi, hatching is the more common method of decoration and mainly around the shoulder in the form of a band or loop. Oblique or vertical lines and in some instances crossed, double or multiple lines have the highest frequencies. At Kilwa, it is hatching or comb or shell impressed patterns that either forms the horizontal or looped bands. (Fig.5.53).The other types of attributes remain almost the same at all the sites.

111

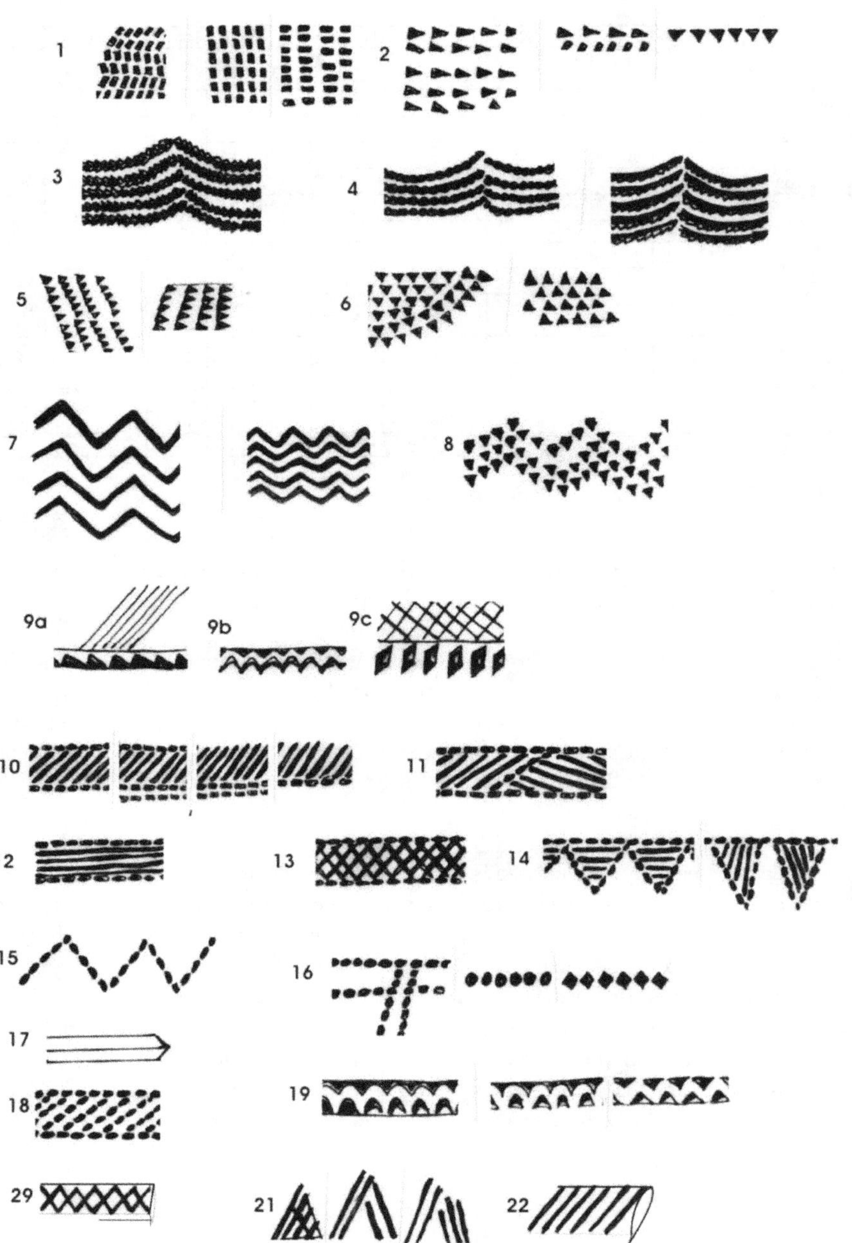

Fig.6.01 (a): Design motifs of different traditions

112

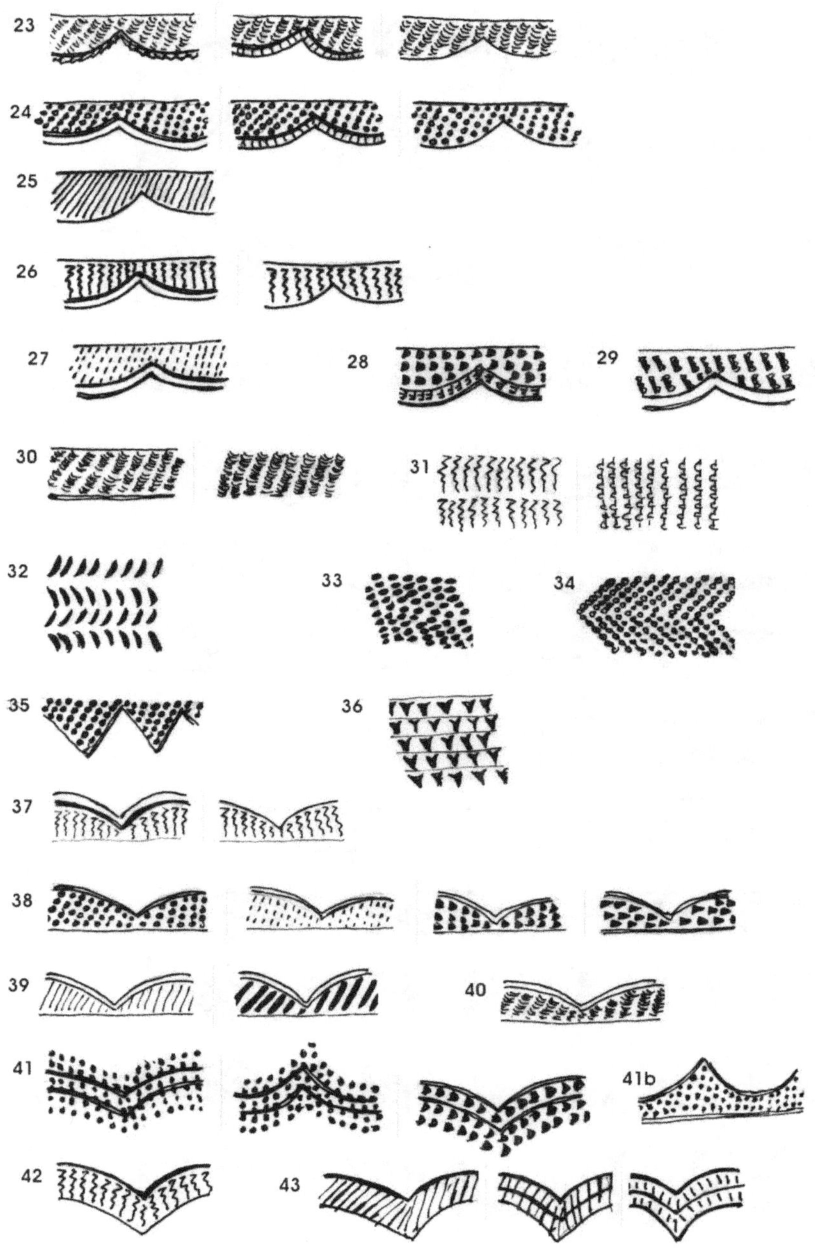

Fig.6.01 (b): Design motifs of different traditions

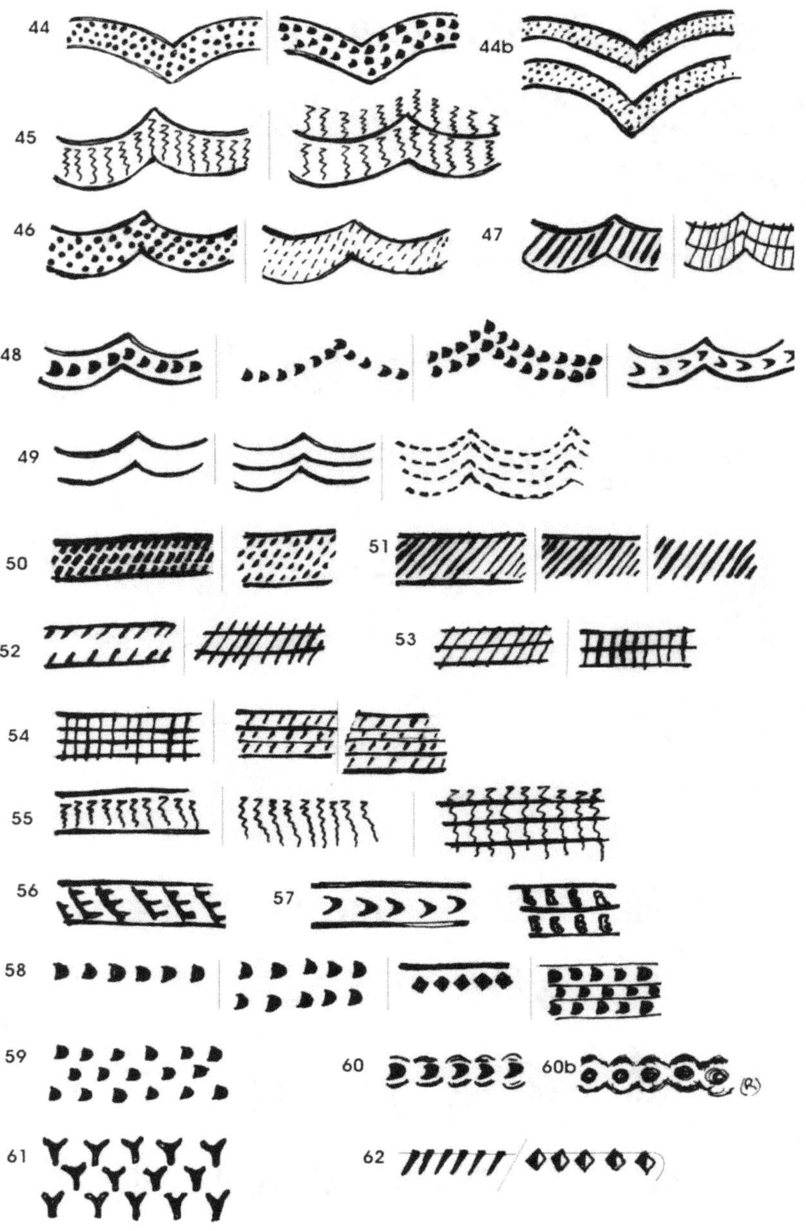

Fig.6.01 (c): Design motifs of different traditions

114

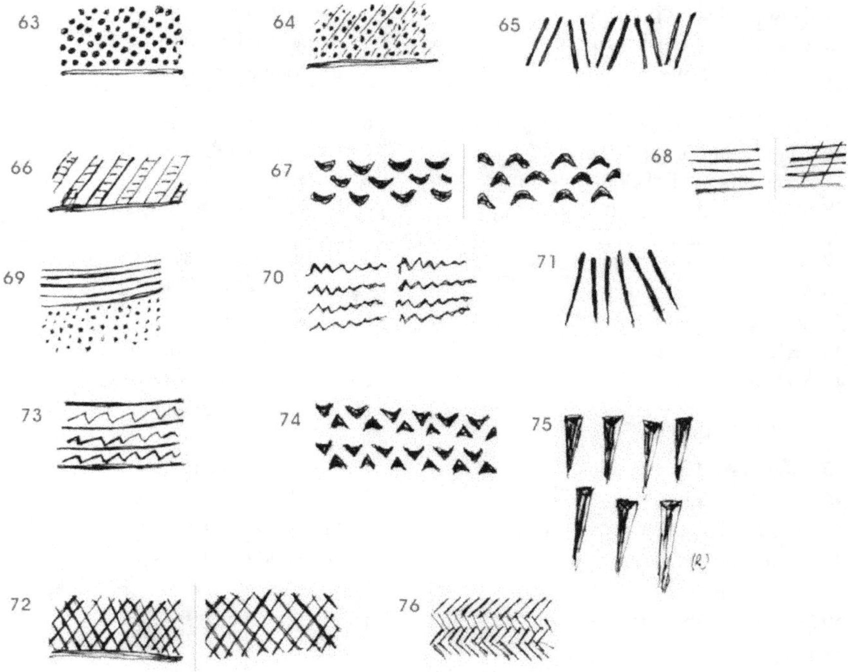

Fig.6.01 (d): Design motifs of different traditions:

Motif/design descriptions for figure 6.01

1 Rectangular stamping	16 Double stamps
2,5,6,8 Varieties of cuneiform stamps	17 Bevels
3-4 Rocker stamping	18 A rim band of dotted lines
7 Zigzag/wavy lines	19 EIW false relief chevrons
9a Oblique hatching and cuneiforms	20 Narrow lip cross hatching
9b False relief chevrons	21 Hatched triangular motifs
9c Cross-hatching and lozenge stamps	22 Lip oblique hatching
10 Band of oblique lines	23 Roulette festoon
11 Interlocking bands of oblique lines	24 Comb stamps festoon
12 Band of horizontal lines	25 Oblique hatched festoon
13 Band of criss-cross lines	26 Impressed shell festoon
14 Pendants	27 Dotted line festoon
15 Zig-zag lines	28 Crescent festoon
29 Stamped festoon	53 Hatched twin-ladders
30 Roulettes band	54 Hatched crossed lines

31 Band of shell impressions	55 Band of shell impressions
32 Rows of finger-nails impressions	56 Band of rake stamps
33-34 Herring bone stamping	57-59 Bounded crescents
35 Festooned chevron with oblique stamps	60a Rocked punctuates
36 Rows of impressed V's	60b Applied knobs
37 Arched band of shell impressions	61 Alternating V's
38 Arched band of stamps	62 Lip indentations
39 Arched band of oblique hatches	63-64 Rim-hanging oblique stamps
40 Arched band of roulettes	65 Mirrored oblique double lines
41 Looped band of stamps	66 Oblique ladders
42 Arched loop of shell impressions	67 Alternating rows of crescents
43 Arched loop of hatched lines	68 Closely hatched horizontal lines
44 Arched loop(s) of stamps	69 Horizontal lines and vertical dotted lines
45 Looped shell impressions	70 Horizontal shell impressions
46 Looped oblique stamps	71 Shoulder wealed vertical rays
47 Looped oblique hatched band	72 Fine neck-criss-cross
48 Looped row of punctuates	73 Bounded chevrons
49 Looped lines	74 Alternating rows of corner stamps
50 Band of oblique set of stamps	75 Moulded wedges
51 Band of oblique set of lines	76 Alternating bands of fine oblique lines
52 Hatched ladder	

Site name	Incis	%	Punct	%	Stamp	%	Graphite	%	Glaze	%	Bevel	%	Total
Mnaida 1	1	50	0		1	50	0		0		0		2
Mnaida 2	9	18.75	19	39.58	12	25	8	16.67	0		0		48
Mnaida 3	14	11.47	55	45.08	35	28.69	17	13.93	1	0.82	0		122
Nakatumbatu	3	100	0		0		0		0		0		3
Kabisela	5	12.19	19	46.34	9	21.95	3	7.32	5	12.2	0		41
Chikayowa	62	22.3	55	19.78	126	45.32	35	12.59	0		0		278
Pemba	41	67.21	0		8	13.11	1	1.64	7	11.46	4	6.56	61
Kitere	3	23.08	8	61.54	2	15.38	0		0		0		13
Mnangole	88	13.86	159	25.04	361	56.85	169	26.61	4	0.63	23	3.62	635
Rushungi	88	74.58	9	7.63	5	4.24	23	19.56	0		0		118
Masakasa	37	39.36	10	10.64	19	20.21	12	12.76	7	7.45	9	9.57	94
Nguruni	11	40.74	4	14.81	0		6	22.22	6	22.22	0		27
Total	362	22.47	338	20.18	578	35.88	267	16.57	30	1.86	36	2.23	1611

Table 6.03: Decorating techniques on excavation findings

Abbr: (Incis) – Incision; (Punct) – Punctations;

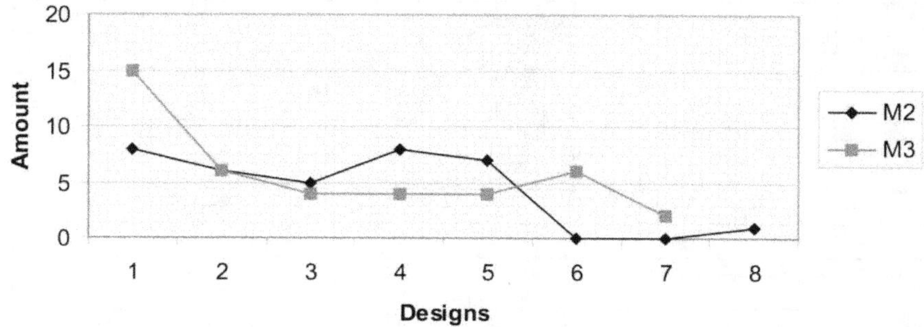

Fig. 6.02: Frequencies of PIW motifs at Mnaida 2 (M2) and Mnaida 3 (M3) compared (the design types 1-8 are shown in fig. 6.01).

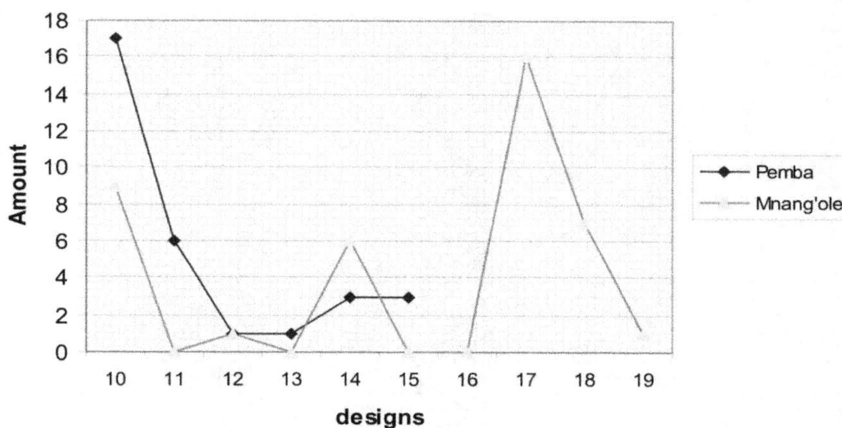

Fig. 6.03: Frequencies of EIW motifs at Pemba and Mnangole compared. The design types 10-19 are shown in figure 6.01).

Motif	Site 1	2	3	4	5	6	7	8	9	10	11	12
1		8	15					1				
2		6	6					0				
3		5	4					1				
4	1	8	4					1			?	
5		7	4					1				
6		0	6									
7		0	2								1	
8		1										
9								1				
10							17		9		3	
11							6		0			
12							1		1			
13							1		0			
14							3		6			
15							3		0			
16									0			
17									16		8	
18									7		0	
19									1		2	
20				2							3	
21				1		2						
22											2	
23				3		10						
24						35			3			
25						1			2		0	
26						2			19		1	
27						1						
28						1						
29						1			1			
30				1		9			3			
31				5		7				1		
32						1						
33						36		6	8			
34						1			0			
35						1			2			
36						0			1			
37						11		5	52			
38			6			4		3	48			
39			0			0			18			
40			0			1			6			
41			3			8		1	0			
42			2			0			32			
43			3			13			1			
44			10			2			2			
45			4			6			103			
46			1			8			7	1	3	
47			0			14			6	7	0	
48			2			4			2	0	1	
49						19			1	2	0	
50		0				1			13	3	5	3
51		4	2		2	7			39	46	7	2
52		0				0			3	2		
53						3			12	10		
54						3			2	3		
55			3			17		4	58	1	5	
56						0			9		1	
57						3			1		0	
58			2		10	7	2	1	5		2	
59						7			0			
60	1		1			1			0			
61						3			1			
62						29		1	33	37	2	1
63			1					2		4		
64			1	1		1			1			
65			1	1	1							
66			1									
67				2		3			1		1	1
68						3			3			
69												
70						1						
71				7		4					10	4
72				0			5				2	
73				1			1					
74				3								
75				1								
76										1	1	

SITES

1. Mnaida 1
2. Mnaida 2
3. Mnaida 3
4. Kabisela
5. Nakatumbatu
6. Chikayowa
7. Pemba
8. Kitere
9. Mnang'ole
10. Rushungi
11. Masakasa
12. Nguruni

Table 6.04: Distribution of Pottery-Designs on different sites given in amount

118

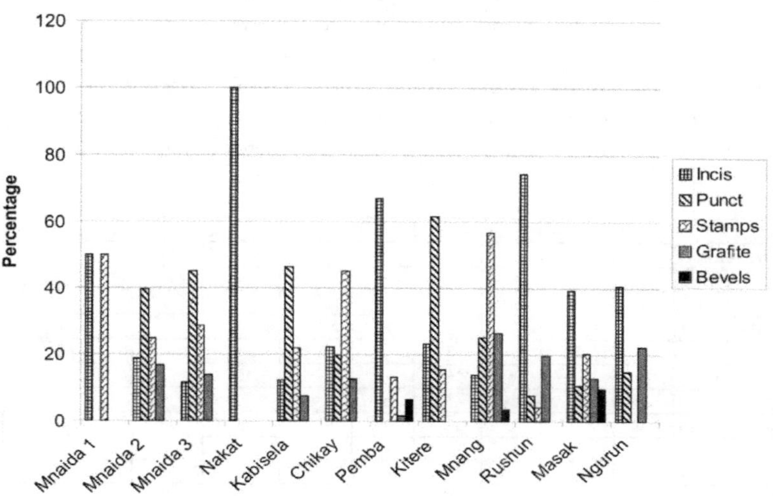

Fig.6.05 (a): Frequencies of diagnostic, decorated, and body sherds at different sites.

Shapes/forms

Several models are used for determining the shapes of pottery vessels. Shepard (1971) developed one model that categorises two groups of vessel forms of restricted and unrestricted shapes. In eastern and southern Africa, another slightly different model was developed from pottery reconstruction (e.g. Soper 1971). This adopted shape classes like necked pots, open bowl, narrow-mouth globular vessel, carinated forms, etc. Huffman (1980), and Collett and Robertshaw (1983) adopted model similar to this in their analyses.

It has been noted that the analytical results obtained from the use of the two mentioned models are more or less the same (Chami 1994). However, problems in working with the first model, especially in differentiating between simple unrestricted vessel forms and unrestricted rim or dependent composite forms were experienced in this work. With smaller fragments and a wide range of rim forms like those flared rims in Plain ware and PSW, or short and only slightly curving rims in TIW, some of the slightly restricted forms could be mistaken for unrestricted shapes.

The second model of reconstructing a range of shapes was adopted herein. As it appears in Table 6.01, a percentage of indeterminable body sherds exist in all the sites, but the range of shapes shown may not reflect the true range of shapes in those groups. A range of narrow-mouth globular vessels and various open bowls and necked pots types are shown in each tradition (Fig. 6.07). Based on the above reason, it was found unrealistic to quantify the percentage of each shape category while leaving behind such a large amount of associated sherds in the indeterminable group.

Site name	TKR	%	TNR	%	NTR	%	IDL	%	Total
Mnaida 1	0		1	50	1	50	0		2
Mnaida 2	0		13	25	38	73.1	1	1.9	52
Mnaida 3	0		14	35	26	65	0		40
Nakatumbatu	0		51	86.4	8	13.6	0		59
Kabisela	2	9.1	3	13.6	15	68.2	2	9.1	22
Chikayowa	0		83	36.2	84	36.7	62	27.1	229
Pemba	2	22.2	0		7	77.8	0		9
Kitere	1	14.3	3	42.8	3	42.8	0		7
Mnang'ole	36	10	149	41.4	69	19.2	106	29.4	360
Rushungi	0		87	58	16	10.7	47	31.3	150
Masakasa	12	9.8	43	34.9	60	48.8	8	6.5	123
Nguruni	5	10.6	20	42.6	22	46.8	0		47
Total	58	5.3	467	42.5	349	31.7	226	20.5	1100

Table 6.05: Rim forms variation

Abbr: TKR= Thickened rim; TNR= thinned rim; NTR= neutral rim; IDL= Indented lip

In Table 6.05, the general characteristic shapes of rims in all the excavated sites are shown. A large percent of all identified rim pieces showed tapered rims, which include a lot of Plain ware rim-sherds from Nakatumbatu, PSW from Rushungi, Mnangole and Chikayowa. Neutral rims that maintain the thickness of the vessel wall are common in the PIW of Mnaida Hill and also in Mtwara EIW pottery. The Kabisela TIW pottery as well contributed many neutral rims. The least percent was that of thickened rims, mainly from Mnangole and Masakasa at Kilwa. They belong to the EIW vessels.

In rim length, the Plain ware has the longest and flaring rims, followed closely by the PSW ware. The shortest rims are those of EIW, which are also the thickest. The EIW generally has a thick body, thicker than any other tradition (Fig. 6.05 b). It look strange that the EIW assemblages do contain very little amount of vessel shape with wall thickness of a centimetre or less. Although the smallest vessels are found in the

PIW ware, it seems that the thinnest and hence lightest vessels are found in the Plain ware, with the exception of imported glass, glazed, or porcelain ware (Fig. 6.05 c).

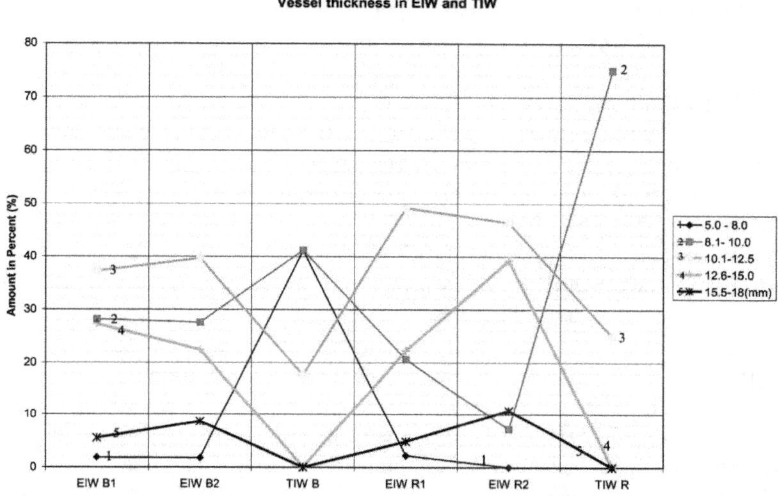

Fig. 6.05(b): Variation in vessel thickness in EIW and TIW. B and R represent Body and Rim thicknesses while 1 is for Mtwara and 2 is for Lindi materials.

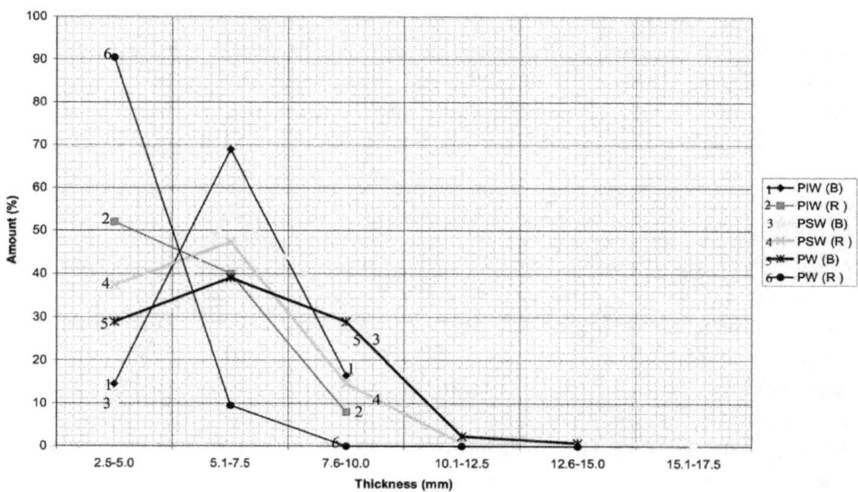

Fig. 6.05(c): Variation in vessel thickness in PIW, PW and PSW. B and R represent Body and Rim thicknesses respectively.

121

The range of forms in the PIW ware is considerably small. Most vessels were relatively thin (Fig. 6.05 c), and comprising small, both shallow and deep bowls and globular pots with slightly or without constricted necks (Fig.6.07: 1a and 1b). There were fewer necked pots than the globular forms or open bowls, some of which could have been preferred for cooking rather than the necked forms.

The EIW ware consists mainly of vessels with constricted necks and relatively short rims. The body wall is generally thick, with a further thickened or neutral rim (Fig. 6.05 b). On the rim end is a series of one or two blunt bevels that carry a band of decorations in the case of Mikindani and partly Mnangole sites. Some of the Mnangole and almost all of the Kilwa EIW pottery have mainly three narrow but clear bevels. Some of the Mikindani EIW rims are very thick and without sharp corners to indicate bevels. Bowl shapes consist of up-turned rims that meet the body at an obtuse angle (Fig. 6.07: 2b, A&B).

The TIW is represented by mainly baggy pots usually with considerably constriction at the neck, with the body relatively thicker than the later traditions, but not very different from the EIW (Fig. 6.05 b; 6.07:3a). The rims are either rounded or slightly thicker with one or two bevels on their outer side to carry decoration (Fig.5.52, 5.36: d-f). In the case of one bevel, it is usually set at an acute angle that makes the rim to appear pointed.

The Plain ware is similarly baggy, but with relatively long and flared rims (Fig.6.07: 4a). The body wall is usually thin (Fig. 6.05 c), thinner than almost all other traditions and mainly with coarse texture. The outer surface is strangely rough and in all sites where it appeared it was in a badly weathered condition. Some of the bowl types have flat bases but others had round bases (Fig. 6.07: 4b). The rims are normally tapered from both sides and with a round tip. In almost all the cases they are not decorated at all.

The last group under discussion is the PSW. It contains open and necked pots, some with necks usually displaying a marked recessed curve in relation to that of the body of the pot. This curve is sometimes so exaggerated that the base of the rim is below the height of the shoulder (Fig; 6.07: 5a, A). Most of the forms with constricted necks usually have indented/ denticulate lips (Fig. 5.47). The bowl-forms of either type rarely bear this feature. Bowls are common, usually of more-or-less hemispherical section with a round base. The rims are mainly tapering and the body is comparatively thin (Fig. 6.05 c). In some instances they have brightly burnished graphite on either side. Necked vessels with double curving profile (Fig. 6.07:5a, 5.48) were also found at Mnangole and Kilwa. A similar type of shape had been previously reported from 'Sanje Kati' site as well as from the Kilwa Island (Chittick 1974: Fig.104).

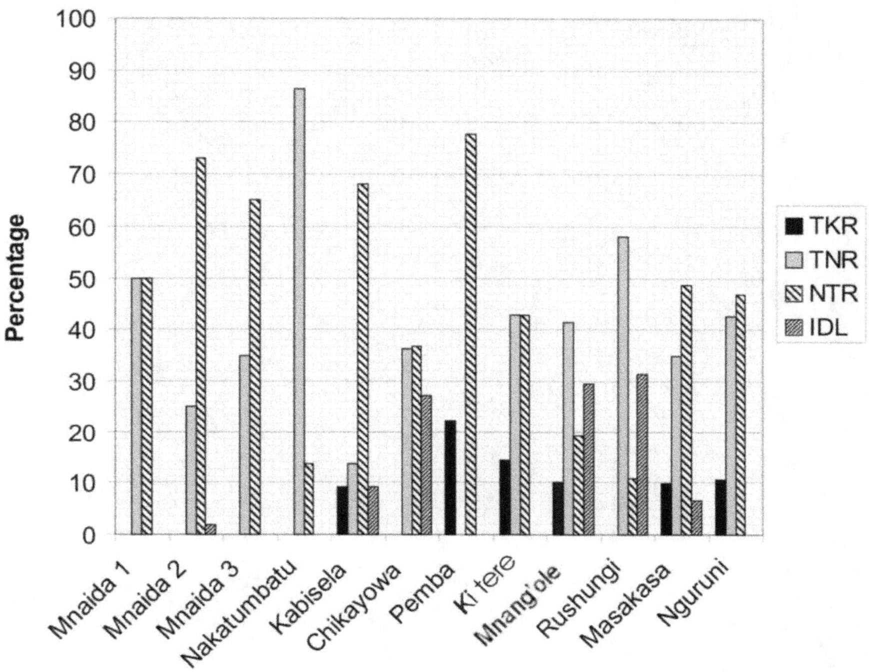

Fig. 6.06: Inter-site rim-form variations: the ratio of Thickened rim (TKR), thinned rim (TNR), neutral rim (NTR), and indented rim (IDL) on different sites is compared.

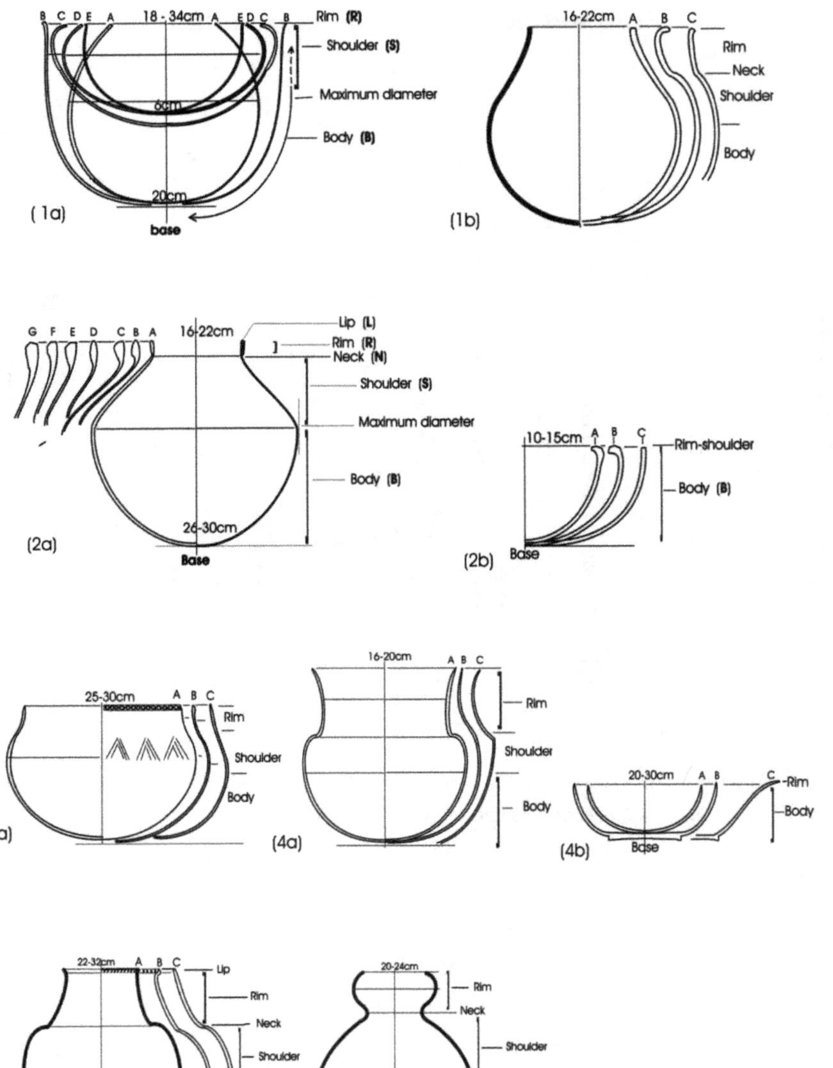

Fig. 6.07: Comparison series of vessel forms: PIW (1a-b), EIW (2a-b), TIW (3a), PW (4a-b), PSW (5a-c).

124

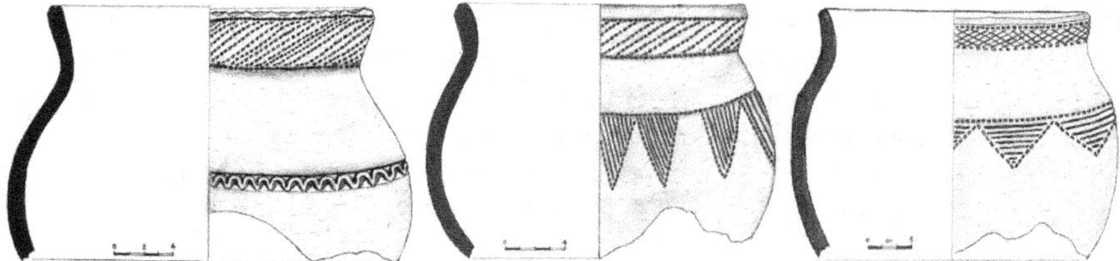

Fig. 6.08: Comparison of EIW pots from Mnang'ole, (left and centre) and Pemba, Mtwara (right)

Lithic Analysis

Stone assemblages studied here are predominantly of quartz (83.3%), although there are small proportions of chert (12%) and quartzite (4.7%). Although such archaeological collections are ubiquitous in the archaeological record, they are rarely studied because the associated analytical problems are so considerable. Mehlman (1989) discussed some of these problems including the difficulty of observing flake properties or retouch, and uncertainties concerning artefact breakage. A hasty study of quartz may result in dubious observations and measurements.

The optical properties of quartz make casual observation of features almost impossible. One of the outcomes is that surface irregularities may be mistaken for retouch or other edge modification, resulting in a spurious assignment to tool type. This is what Glynn Isaac used to refer to as "over-typing" (in Mehlman 1989: 118). Based on this situation, this work tried to ensure that retouch on quartz tools was not a product of subjective imagination. In this study most of the stone artefacts were examined under a binocular microscope to verify any retouch or surface modification.

The converse of an over-typing problem is what Mehlman (1989) termed "under-typing". In under-typing case, there is a tendency of not inspecting each individual piece because of the above-mentioned optical problems, and hence focusing on a few and most obviously shaped tools. Currently, there are few alternative ways for describing new assemblages that have taken into account the embodied problems of quartz material. One may choose to employ extant typologies, in which case the dilemma is which one to select. Alternatively, one may develop yet another typology, probably relying on the others in the formulation of certain types. This work preferred to use Mehlman's typological scheme, not only because it combined other popular typologies in East African lithic studies, but also because of its similarity to the quartz artefacts involved. In addition, the collection is not large and diversified enough to develop a new scheme.

However, it has been argued that the application of a typology tends to blind the analyst to an understanding of manufacturing techniques (Sampson 1979). There are also an infinity number of attributes that one could measure in any stone artefacts, but most lithic analysts routinely measure attributes that are believed to be relevant to the understanding of the research questions. This work is one of the research examples with specific questions to answer, the major question being about the evidence, which suggests the presence of early, settled communities on the southern coast. Only the Masakasa and Kitere sites provided most of the lithic components during excavations, and the third site, Mbwemkuru, showed MSA tools during the survey.

	I: Retouched pieces	%		II: Cores (continued)		%
A	**Scrapers: (11)**	*1.07*	D	**Bipolar cores:**		
	convex-end scraper 1			Bipolar cores 37		*3.61*
	convex end/side scraper 4	*0.39*				
	circular scraper 2	*0.19*	E	**Other cores:**		
	convex side scraper 3	*0.29*		Amorphous/casual cores	5	*0.49*
	sundry double end/side scraper 1			core fragments	38	*3.71*
B	**Backed pieces: (4)**	*0.39*		**III: Debitage**		*85.17*
	curve-backed pieces 4		A	**Angular fragments: (79)**		*7.71*
				angular fragments	58	*5.66*
C	**Points: (4)**	*0.39*		blade segments	21	2.05
	Unifacial points 2					
	Bifacial point 2		B	**Specialized flakes:**		
				burin spall	3	*0.29*
	II: Cores (94)	*9.17*				
A	**Peripherally worked:**		C	**Flakes: (708)**		*69.07*
	Disc core 1			whole flakes	647	*63.12*
				Utilized/trimmed flake	3	*0.29*
B	**Patterned platform: (50)**	*4.88*		flake talon fragments	58	*5.66*
	Primatic cores 1					
	Single platform cores 37	*3.61*	D	**Blades: (83)**		*8.10*
	Double opposed platform core 4	*0.39*		whole blades	62	*6.05*
	Double adjacent platform core 8	*0.78*		Utilized/trimmed blades	1	
				Blade talon fragments	19	*1.85*
C	**Intermediate cores:**			utilized/trimmed blade talon	1	
	Platform/bipolar cores 2	*0.20*				

Table 6.06: Typological summary of the stone artefacts based on Mehlman's scheme

The Masakasa assemblage consisted of 86.9% Quartz, 8.8% Chert and only 4.3% Quartzite (Fig.6.09). At Kitere 33.96% was Quartz, 56.6% Chert and 9.4% Quartzite. At Mnang'ole excavations 86.7% of the relatively few lithic artefacts were of Quartz, and 13.3 % of Chert. From Mbwemkuru only 9 tools were collected during survey, eight of them made of Chert rock (Table 6.07). Putting it all together, 96% of Quartz artefacts (n=821) came from Masakasa while 2.1 % (n=18) was from Kitere. Most of the Quartzite artefacts (85.4%) were also from Masakasa (n=41) and 10.5% from Kitere (n=5). Masakasa provided 67.5% of the Chert artefacts while Kitere contributed 24.4% (n=30) and Mbwemkuru another 7.3%.

Mehl #	Artifact name	2a	2b	2c	2	3a	3b	3c	3	8a	8b	8c	8	9a	9b	9c	9	11a	11b	11c	11	13a	13b	13c	13	Total	%
2	Convex end scraper																		1		1					1	
4	Cvx end/side scraper																							1	3	4	4	0.39
5	Circular csraper																			1		1			1	1	2	0.19
7	Convex side scraper													1			1	2			2					3	0.29
12	Sundry db-end/side scr																								1	1	1	
27	Curve-backed piece						1		1	1			1					1		1	2					4	0.39
35	Unifacial point																		1		1			1	1	2	0.19
37	Bifacial point																							2	2	2	0.19
59	Disc core																			1	1					1	
61	Prismatic core																	1			1					1	
62	Single ptatform core																	36		1	37					37	3.61
64	Double opposed p-core																	4			4					4	0.39
66	Double adjacent p-core																	7		1	8					8	0.78
71	Platform/bipolar core																	2			2					2	0.19
74	Bipolar core																	37			37					37	3.61
76	Amorphous core																	5			5					5	0.49
77	Core fragments									1	2	3	6	1			1	31			31					38	3.71
78	Angular fragments									5	1	8	14	7			7	37			37					58	5.66
80	Blade segments													2		1	3	16	2		18					21	2.05
82	Burin spall																	3			3					3	0.29
84	Whole flake	2			2					11		11	22	1		1	2	526	28	67	621					647	63.12
85	Utilized (T/U) flake																	2		1	3					3	0.29
86	Flake talon fragment											6	6					48	1	3	52					58	5.66
88	Whole blade											2	2	1			1	48	5	6	59					62	6.05
89	Utilized (T/U) blade																	1			1					1	
90	Blade talon fragment										2		2					13	2	2	17					19	1.85
91	Utilized blade-talon frag																	1			1					1	
	Raw-Material range	2					1			18	5	30		13		2		821	41	83			1	8		1025	
	Site-based Total				2				1				53				15				945				9	1025	
	Smaller chips				4				1				41				25				908						
	Total				6				2				94				40				1853				9	2004	

Table 6.07: Stone artefact categories based on raw material type

Abbr: Raw Material (1st row): - Quartz (a), Quartzite (b), Chert (c). Sites: - Mnaida 2 (2), Mnaida 3 (3), Kitere (8), Mnangole (9), Masakasa (11), Upper Mbwemkuru (13)

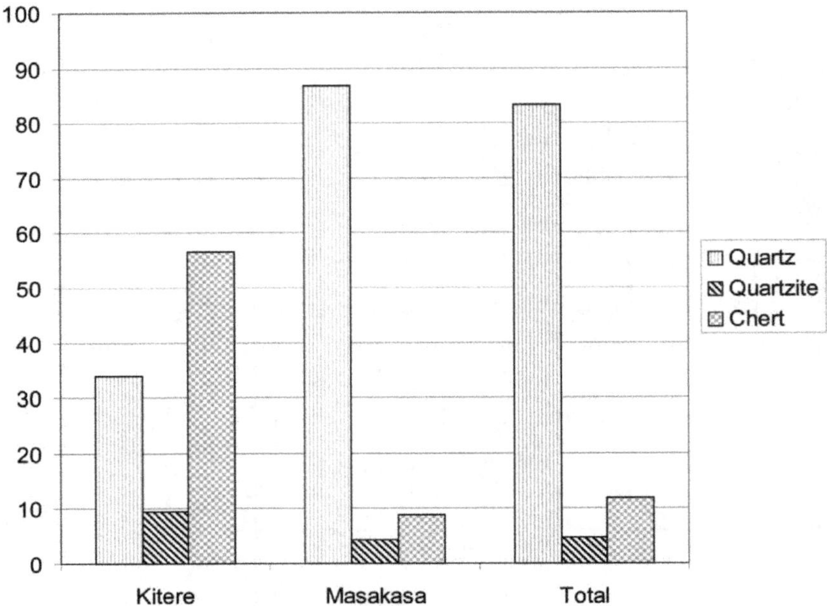

Fig. 6.09: General preference (total) and specific site preference for particular raw materials compared.

The vertical distribution of stone artefacts in this work seems to differ with sites. There have been associations of lithics and pottery artefacts in most of the sites. At Mnaida the lithics were collected below pottery horizon. At Kitere site, pottery and lithics coexisted especially in the 4th and 5th levels but below these levels was lithic only. At Mnangole, most artefacts were in association with the EIW pottery. Masakasa site provided the best evidence of pre-ceramic stone working tradition. Between levels 8 and 13 collected artefacts were only lithics. There is a considerable overlapping of the pottery horizons, but maximum lithic concentrations were found below pottery levels, especially in levels 8-11 of trench 2, levels 9-10 of trench 3, levels 6-7 of trench1 and in level 9 of trench 4 (Table 6.08). As already argued in this work, this variation in vertical distribution may have very little or insignificant temporal meaning, a consequence of original surface topography during the formation of the site.

128

Site	2a	2b	2c	2	3a	3b	3c	3	8a	8b	8c	8	9a	9b	9c	9	11a	11b	11c	11	13a	13b	13c	13	Total
Depth																									
Surface					0	1	0	1									6	6	3	15	0	1	8	9	25
30 - 40 cm									7	2	10	19					14	0	1	15					34
40 - 50 cm	2	0	0	2					8	2	13	23	1	0	0	1	58	7	4	69					95
50 - 60 cm									3	0	10	13	4	0	1	5	74	3	4	81					99
60 - 70 cm													4	0	1	5	81	4	4	89					94
70 - 80 cm													4	0	0	4	115	5	20	140					144
80 - 90 cm																	257	5	21	283					283
90 - 100 cm																	129	6	17	152					152
100 - 110 cm																	46	3	0	49					49
110 - 120 cm																	21	1	6	28					28
120 - 130 cm																	19	1	2	22					22
Total	2	0	0	2	0	1	0	1	18	4	33	55	13	0	2	15	820	41	82	943	0	1	8	9	1025

Raw material																									Total
Quartz				2				0				18				13				820				0	853
Quartzite				0				1				4				0				41				1	47
Chert/Flint				0				0				33				2				82				8	125
																								Total	1025

Table 6.08:Statigraphic distribution of stone artefacts and their raw material types.

Abbr: Raw Material (1st row): - Quartz (a), Quartzite (b), Chert (c). Sites: - Mnaida 2 (2), Mnaida 3 (3), Kitere (8), Mnangole (9), Masakasa (11), Upper Mbwemkuru (13).

In examining the surface features of the artefacts, it was learnt that there was high proportion of end struck pieces, 88.6%. This examination excluded 10.3% of the artefacts without visible platform (Fig.6.10). The high frequency of the end struck pieces might be reflecting the smaller size of the cores, in that the maker was trying to maximise the length of the flake by striking perpendicular to the longest axis of the core.

Attribute	Site	2a	2b	2c	2	3a	3b	3c	3	8a	8b	8c	8	9a	9b	9c	9	11a	11b	11c	11	13a	13b	13c	13	Total
Morphology	with platform	2			2	1			1	12	2	24	38	3	0	1	4	745	39	78	862		1	8	9	916
	with-no platform	0			0	0			0	6	3	9	18	10	0	1	11	70	2	4	76					105
	End strike	1			1	0			0	7	0	19	26	3	0	1	4	694	34	68	796		1	8	9	836
Strike-type	Side strike	1			1	1			1	5	1	5	11					75	6	14	95		0	0	0	108
	No cortex	1			1	1			1	5	4	18	27	8	0	2	10	268	28	58	354	0	1	5	6	399
	100% cortex									1	0	5	6	0	0	0	0	124	3	2	129	0	0	0	0	135
Cortex	75% cortex	1			1					1	0	2	3	0	0	0	0	107	1	3	111	0	0	0	0	115
	50% cortex									2	0	2	4	1	0	0	1	112	2	4	118	0	0	1	1	124
	<50% cortex									9	2	6	17	4	0	0	4	204	6	15	225	0	0	2	2	248
Fracture end	Indeterm									6	4	13	23	8	0	1	9	196	2	7	205					237
	Feather	2			2	1			1	7	0	14	21	3	0	0	3	386	26	57	469		1	8	9	505
	Step									1	0	3	4	0	0	0	0	35	3	2	40					44
	Hinge									0	0	0	0	0	0	0	0	24	1	6	31					31
	Plunging									4	0	3	7	2	0	1	3	174	7	10	191					201
Core-discard	Early											1	1					17	0	0	17					(18)14.1%
	Premature											1	1					28	0	1	29					(30) 23.4%
	Exhausted										2	2	4	1			1	74	0	1	75					(80) 62.5%
	Sample size				6				2				94				40				1853				9	2004

Table 6.09:Frequencies of attributes of artefacts in relation to the raw material.

Abbr: **Raw Material** (1st row): - Quartz (a), Quartzite (b), Chert (c). **Sites:** - Mnaida 2 (2), Mnaida 3 (3), Kitere (8), Mnangole (9), Masakasa (11), Upper Mbwemkuru (13).

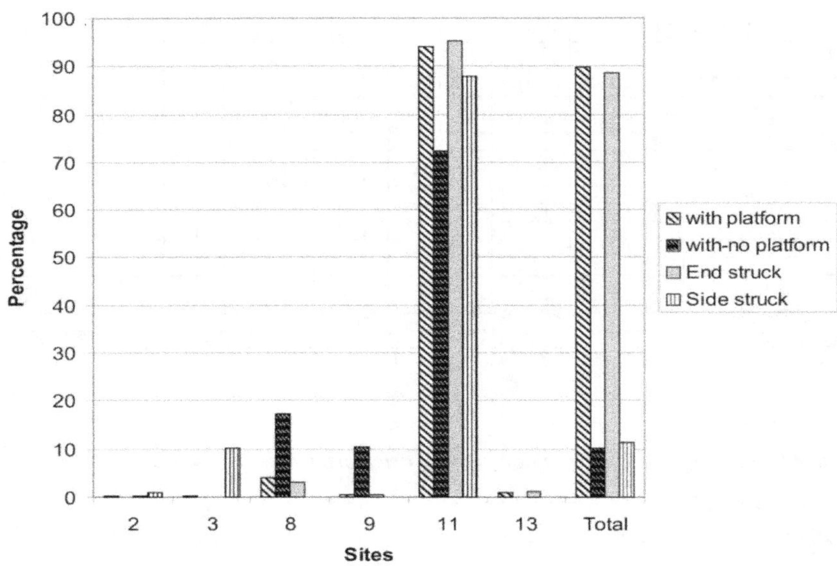

Fig.6.10: Morphological features of stone artefacts compared: Mnaida 2 (2), Mnaida 3 (3), Kitere (8), Mnangole (9), Masakasa (11), and Upper Mbwemkuru (13).

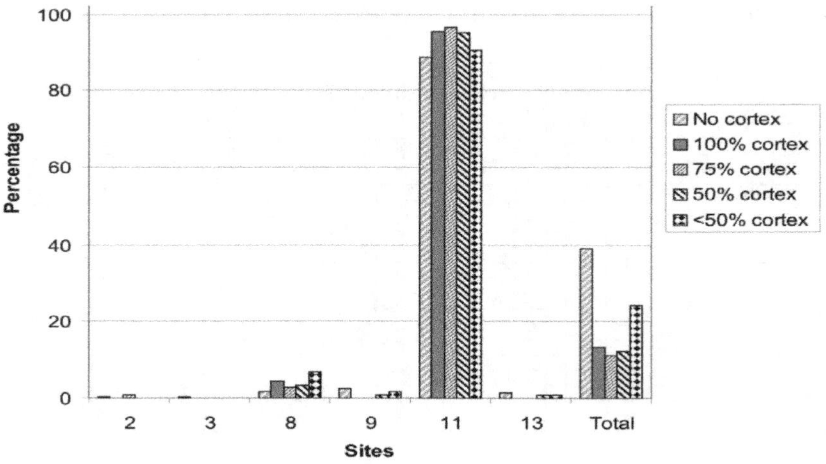

Fig. 6.11: Amount of remaining cortex on the artefact surface compared: (2) Mnaida 2; (3) Mnaida 3; (8) Kitere; (9) Mnang'ole; (11) Masakasa; (13) Mbwemkuru.

130

Most of the lithic artefacts (61%) had a varying degree of cortex (Fig. 6.11). This figure excluded 908 smallest angular fragments that could not be measured. The flake termination is mainly of feather fracture or plunging termination. The size of the cores used in making the stone artefacts was very small (Fig. 6.14). Most of the recovered cores suggested abandonment at exhausted stage (Table 6.09). This is probably because of its smaller size, in such that they became worthless after just few struck. The survey of the region could not discover source of Quartz pebbles suitable for the production of larger pieces. The makers seem to have relied on the local source only, which in turn contained only small alluvial pebbles.

The amount of cores or core fragments counted to only 9.2% of the stone artefacts. Most cores are either single platform or bipolar cores (Fig. 6.13). Because of smaller size of the cores, it should have been much easier to work on an anvil during flaking. It is very unfortunate that the number of finished tools is too small (ca. 2%) to make any comment on the technology.

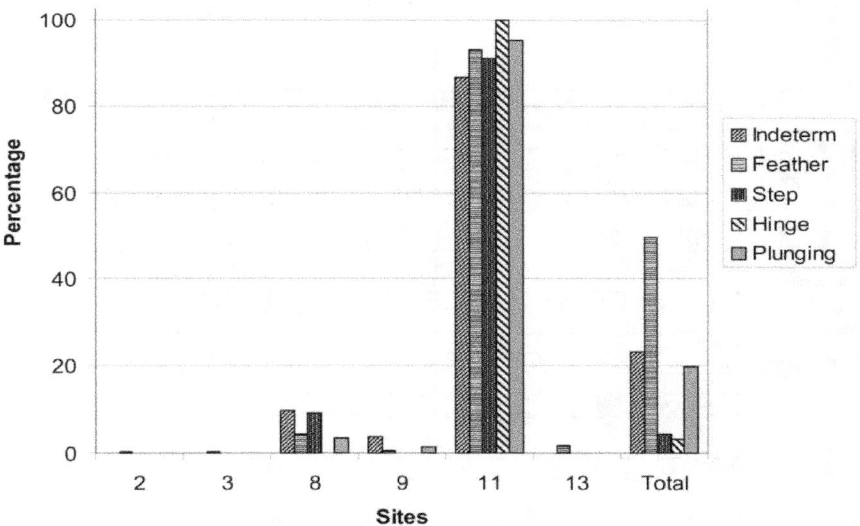

Fig.6.12: Flake-end characteristics of all stone artefacts compared. : (2) Mnaida 2; (3) Mnaida 3; (8) Kitere; (9) Mnang'ole; (11) Masakasa; (13) Mbwemkuru.

Attribute	Site	2a	2b	2c	2	3a	3b	3c	3	8a	8b	8c	8	9a	9b	9c	9	11a	11b	11c	11	13a	13b	13c	13	Total
Retouch-position	Direct																	4	3	1	8		1	4	5	13
	Bifacial																						0	4	4	4
	Alternate																									
	Inverse													1			1									1
	Alternating						1		1																	1
Retouch-extent	Marginal						1		1					1			1	3	0	1	4		0	0	0	6
	Invasive																	0	3	0	3		2	3	5	8
	Covering																	1	0	0	1		0	3	3	4
	Marg/invas																						0	0	0	0
	Marg/covering																						0	2	2	2
	Invas/covering																						0	0	0	0
Material	Quartz								1								1				4				0	6
	Quartzite								0								0				3				1	4
	Chert/flint								0								0				1				8	9
	Total				0				1				0				1				8				9	19
	Sample size				6				2				94				40				1853				9	2004

Table 6.10: Secondary modifications on stone tools in relation to the raw material

Abbr: **Raw Material** (1st row): - Quartz (a), Quartzite (b), Chert (c). **Sites**: - Mnaida 2 (2), Mnaida 3 (3), Kitere (8), Mnangole (9), Masakasa (11), Upper Mbwemkuru (13).

Attribute	Artifact type	78	80	82	84	85	86	88	89	90	91	Total
Length	Minimum	12.0	14.0	15.0	9.0	13.0	9.5	16.3	20.0	13.5	19.0	
	Maximum	61.0	29.0	19.0	70.0	32.0	30.0	53.0	20.0	34.0	19.0	
	Average	**21.2**	**21.3**	**16.5**	**21.8**	**20.8**	**18.2**	**27.9**	**20.0**	**29.9**	**19.0**	
Width	Minimum	8.0	10.0	10.0	6.0	18.6	11.0	6.0	13.0	9.0	13.0	
	Maximum	56.0	24.0	13.0	50.6	30.0	24.5	25.0	13.0	26.0	13.0	
	Average	**14.9**	**16.4**	**11.2**	**16.8**	**24.5**	**17.5**	**12.8**	**13.0**	**15.6**	**13.0**	
Thickness	Minimum	2.0	2.0	6.0	2.0	3.0	2.0	2.0	5.0	2.2	6.0	
	Maximum	25.6	7.0	8.0	18.0	8.0	9.0	10.0	5.0	9.0	6.0	
	Average	**6.4**	**4.5**	**6.7**	**6.0**	**5.0**	**5.2**	**5.1**	**5.0**	**4.9**	**6.0**	
	Sample size	58	21	3	647	3	58	62	1	19	1	873

Table 6.11 :General size-range of the lithic debitage from different sites.

Abbr: (1st row): Artefact type: angular fragments (78), blade segments (80), burin spar (82), whole flake (84), trimmed/utilized (T/U) flakes (85), flake talon fragments (86), whole blade (88), trimmed/utilized (T/U) blade (89), blade talon fragments (90), trimmed/utilized blade talon fragment (91)

Attribute	Core type	59	61	62	64	66	71	74	76	77	Total
Length	Minimum	37.0	18.0	14.0	10.0	19.0	21.0	15.0	21.0	16.0	
	Maximum	37.0	18.0	60.4	26.0	42.0	42.0	36.0	30.0	42.0	
	Average	37.0	18.0	26.3	19.5	26.8	31.5	23.6	24.2	25.2	
Width	Minimum	24.0	28.0	13.0	20.0	12.0	25.0	11.0	17.0	10.0	
	Maximum	24.0	28.0	62.0	30.0	31.0	39.0	44.0	20.0	50.0	
	Average	24.0	28.0	23.7	23.0	21.1	32.0	21.0	18.8	18.7	
Thickness	Minimum	11.0	19.0	6.0	14.0	12.0	18.0	9.0	10.0	4.0	
	Maximum	11.0	19.0	59.0	20.0	30.0	33.5	25.0	17.0	19.0	
	Average	11.0	19.0	17.3	16.4	17.2	25.8	14.7	14.4	10.9	
	Sample size	1	1	37	4	8	2	37	5	38	94

Table 6.12: General core types and size range

Abbr: (Core types): - Disc core (59), Prismatic core (61), single platform core (62), double-opposed platform core (64), double adjacent platform core (66), platform/ bipolar core (71), bipolar core (74), amorphous core (76), core fragments (77).

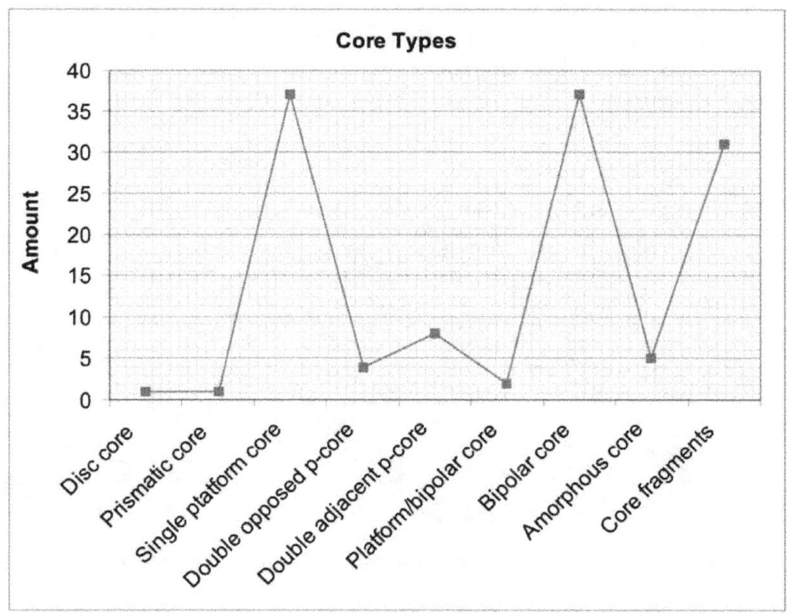

Fig. 6.13: Frequencies of different core types at Masakasa site.

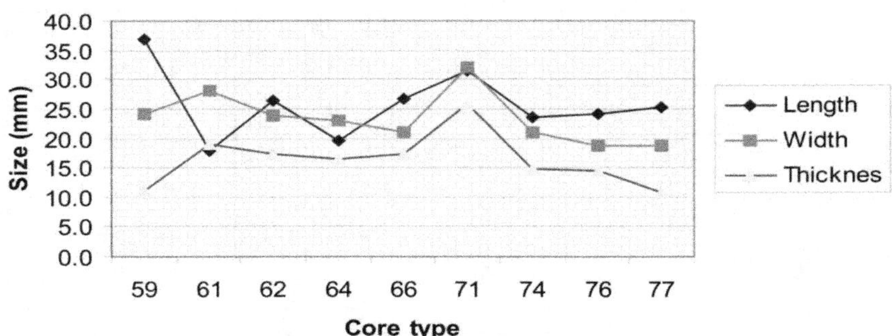

Fig. 6.14: Average core sizes compared: Disc core (59), Prismatic core (61), single platform core (62), double-opposed platform core (64), double adjacent platform core (66), platform/bipolar core (71), bipolar core (74), amorphous core (76), core fragments (77).

Formal tools were quite rare in most of the sites, representing only 2% of lithics artefacts (Table 6.13a). Out of this, 33% are made of Quartz, 24% Quartzite and the rest are Chert (43%). Most (43%) of the formal tools are from Masakasa and Mbwemkuru sites. On the other hand, there is a large amount of smaller angular fragments (rock chips) at Masakasa (n=908) comprising 49% of the total lithics, with Kitere site containing 43.6% of chips in its lithic assemblage (n=41) while at Mnangole the chips was about 62.5% (n=25). All the cores discussed here were collected at Masakasa, albeit a few core fragments recovered at Kitere site.

Mehlman's category	Artifact name	Total	Site 2	Site 3	Site 8	Site 9	Site 11	Site 13	R/Mat 1	R/Mat 2	R/Mat 3	Length		Width		Thickness		Avg X	Avg Y	Avg Z
2	Convex end scraper	**1**					1			1		82.0		57.4		18.0		82.0	57.4	18.0
4	Cvx end/side scraper	**4**						4		1	3	78.0	65.0	60.0	48.0	37.5	17.0	70.6	54.0	23.1
5	circular csraper	**2**					1	1		1	1	75.0	72.0	70.0	61.0	26.0	22.0	73.5	65.5	24.0
7	Convex side scraper	**3**				1	2		3			38.0	29.0	30.5	20.0	8.0	5.0	32.7	23.8	6.5
12	Sundry db-end/side scr	**1**						1			1	91.0		55.0		23.0		91.0	55.0	23.0
27	Curve-backed piece	**4**		1	1		2		2	1	1	40.0	30.0	55.0	13.5	12.0	4.5	35.5	31.4	8.1
35	Unifacial point	**2**					1	1		1	1	80.0	57.0	56.0	52.0	21.5	15.0	68.5	54.0	18.3
37	Bifacial point	**2**						2			2	85.0	60.0	75.0	50.0	29.0	18.5	72.5	62.5	23.8
89	Utilized (T/U) blade	**1**					1		1			20.0		13.0		5.0		20.0	13.0	5.0
91	Utilized blade-talon frag	**1**					1		1			19.0		13.0		6.0		19.0	13.0	6.0
	Site-total			1	1		9	9												
	Raw Material				1. Quartz	7														
					2. Quartzite		5													
					3. Chert/Flint			9												
	Total	21 (2.05% of the lithic artifacts)																		

Table 6.13a: Size range and raw material type of the formal stone tools

Abbr: (**Sites**) - Mnaida 2 (2), Mnaida 3 (3), Kitere (8), Mnangole (9), Masakasa (11), Upper Mbwemkuru (13); (**Raw-materials**) Quartz (1), Quartzite (2), Chert (3).

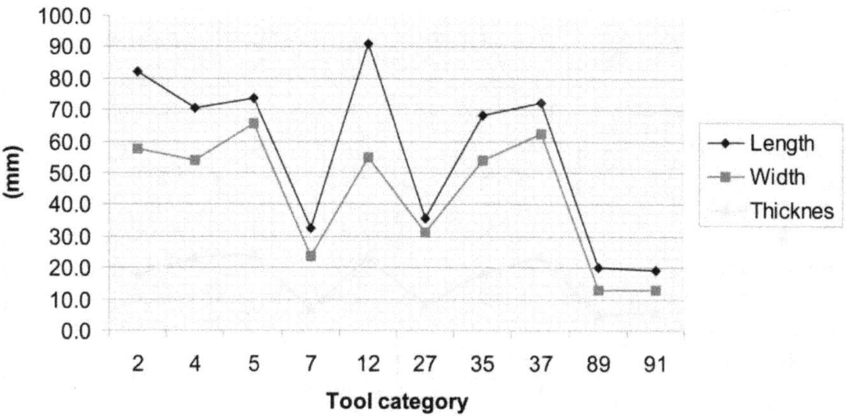

Fig. 6.15: Average tool sizes compared: convex end scraper (2), convex end/side scraper. (4), circular scraper (5), convex-Side scraper (7), sundry db-end/side scraper. (12), curved backed pieces (27), unifacial points (35), bifacial points (37), T/U blades (89), and T/U blade talon fragments (91).

Tool	*Strike*	*R/Mat*
Convex end scraper 1	end-struck	Quartzite
Convex end/side scraper 4	end-struck 4	Quartzite 1, Chert 3
Circular scraper 2	end-struck 2	Quartzite, Chert
Convex side scraper 3	end-struck 3	Quartz
Sundry double-end/side scraper 1	end-struck	Chert
Curve-backed pieces 4	end-struck 3	Quartz 2, Chert 1
	side-struck 1	Quartize
Unifacial point 2	end-struck 2	Quartzite, Chert
Bifacial point 2	end-struck 2	Chert
Utilized/trimmed flakes 3	end-struck 3	Quartz 2, Chert 1
Utilized/trimmed blade 1	end-struck	Quartz
Utiliized blade-talon frag. 1	end-struck	Quartz

Table 6.13 b: Strike-type and raw materials compared

135

Tool	Raw-material	Retouching	Retouch Extent
Convex end scraper 1	Quartzite	Direct	Invasive
Convex end/side scraper 4	Chert	Direct 2,	Covering
	Chert	Bifacial	Invasive
	Quartzite	Direct	Invasive
Circular scraper 2	Quartzite	Direct	Invasive
	Chert	Bifacial	Covering
Convex side scraper 3	Quartz	Direct 2,	Marginal 2
		Inverse	Marginal
Sundry double end/side scraper 1	Chert	Direct	Invasive
Curve-backed pieces 4	Quartz 2	None	Nill
	Quartzite	Alternating	Marginal
	Chert	Direct	Marginal
Unifacial point 2	Quartzite	Direct	Invasive
	Chert	Direct	Invasive
Bifacial point 2	Chert 2	Bifacial	Marginal/covering
Utilized/trimmed flakes 3	Quartz, Chert 2	None	Nill
Utilized/trimmed blade 1	Quartz	Direct	Marginal
Utiliized blade-talon frag. 1	Quartz	Direct	Covering

Table 6.13c: Raw material and retouching characteristics compared

In the above Table 6.13b, the striking directions on different lithic raw material are examined. In a similar way, retouching characteristics are shown in relation to the artefact raw material (Table 6.13c). The observed patterns as shown in these tables are random and unpredictable. There is no clear relationship between the striking position, the retouch type and its extent, with the artefact raw material. The shape of the core and flake edges might have been determining the striking and retouching behaviour of these tools, rather than the desire of the maker or raw material type. The proportion of retouched pieces in general is insignificantly small. The component of backing behaviour in all the sites is almost absent although the amount of suitable pieces that could be backed was considerably large.

Chronology

This section provides the chronological aspects of cultural development and changes along the coast of southern Tanzania. The measurement of time and ordering of prehistoric cultures in chronological sequence has been one of the archaeologists' major preoccupations since the very beginning of scientific researches (Fagan 2005: 134). In this section, the analysis of the chronological relationships among artefacts, sites and other features recorded during the fieldwork is provided.

Archaeologists have relied on two types of chronology, the relative and absolute chronology. In the first type, one establishes chronological relationship between sites and culture through careful observation of sequences of occupational levels and correlates these with cultural sequences at other sites in the same area. The other type refers to dates obtained scientifically. Relative chronology, which is argued to be the foundation of all archaeological research, is established either by stratigraphic observation or by placing artefacts in chronological order (Fagan 2005: 188). It has also been noted that chronological sequence can be horizontal, as when economic or socio-political conditions dictate regular movement of villagers, when the field is exhausted or resident rules are modified. In this case, a cultural sequence may be scattered throughout a series of single-level occupation sites over a large area and can be put together only by judicious survey work and careful analysis of artefacts found in different sites (Fagan 2005: 140).

Chronological 'types' are usually established through analysis of the recovered archaeological record in this case especially pottery. These types are defined in terms of attributes that show changes over time. The relative chronology method has been used to check with both already defined types and existing dates of similar types in eastern and southern Africa. It has been argued in this thesis that cultural sequence is scattered over a large stretch of the coast with some local variations. It is the cultural sequence of the types, rather than the depth at which the material is found, which has been useful in defining the relationships of the sites. Absolute chronology is used in this work to define the ordering of the chronological types that occur as single-level occupations. The two methods have been successfully used to reconstruct the stratigraphic unit of the southern coast that corresponds to ancient documents, as well as to archaeological and linguistic arguments.

General Stratigraphy

The land survey, excavations and analyses of the finds have proved the presence of a long cultural sequence on the Southern coast, dating back probably to the Middle Stone Age. The woodland that extends from the coast to the interior seems to have been occupied by long-established social groups that had been adapting and diversifying for millennia until recently. Occupations and settlements have been shifting around and interactions have taken place over many years. During the analyses of the lithic and pottery artefacts, different cultural phases were identified as described in the following sections.

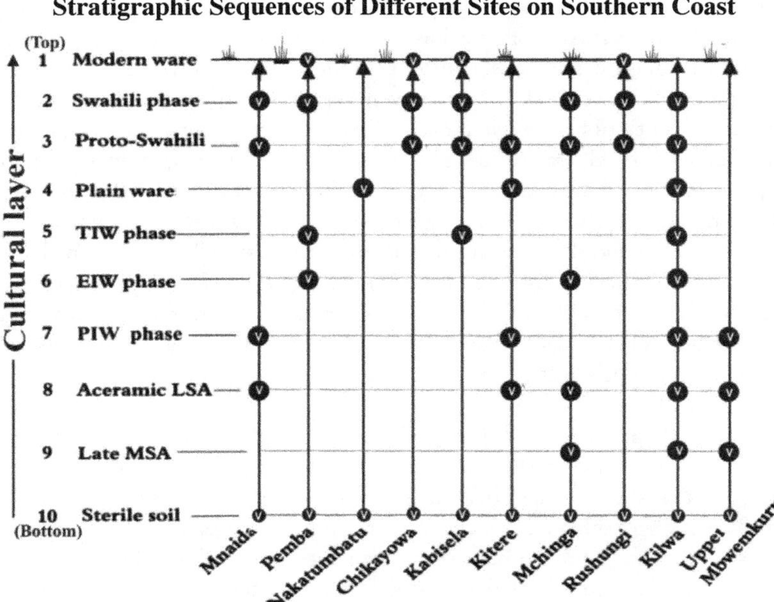

Fig. 6.16: Stratigraphic sequences of different sites on Southern coast of Tanzania (marked).

Paleolithic phases.

This work verified the distribution of MSA- derived lithic tools in three sites. Although excavations and radiometric dating could not fully demonstrate its diachronic pattern, spatial distribution of the tool types was evident at the north-eastern corner of the Kilwa Island, woodland areas along the Mbwemkuru River down to the shore, and the woodland area of Mchinga. The Upper Mbwemkuru is possibly the core for this tradition, with well-fashioned Levalloisian flake tools. This could be the oldest occupation on the coast, but still to be conclusively dated. No similar material was recovered during the excavations, but the upper Mbwemkuru valley could possibly provide a better sequence that seems to date back into Pleistocene times. Based on Sinclair et al's evaluation (1993: 416), there is little doubt that foraging communities were well-established in most parts of northern coast of Mozambique throughout the Holocene, corresponding to what was observed on the adjacent coast of southern Tanzania.

The oldest recorded cultural tradition in the excavations seems to be of the Late Stone Age (LSA). Blades and microlithic artefacts represent this tradition at Kilwa

138

Island, Mnangole/Mchinga, Kitere and Mnaida sites (Fig.6.16). In the Mbwemkuru valley, a good number of these artefacts could be seen in the eroded gullies, mixed with MSA-derived tools. The tradition seems to have continued at some places during the Ironworking period, as evidenced at Kilwa and Mnangole areas.

PIW and EIW phases

The lowest and hence the oldest pottery tradition that appears especially at Mnaida 1 and Mnaida 2, partly at Mnaida 3, at Kitere and Kilwa Island; has been conventionally named as Mnaida tradition. This pottery is quite distinctive, with thin walls and cuneiform or rocker-stamping decorations. At Kilwa, the tradition clearly occurs below the EIW horizon. At Masakasa trench 3, it appears between levels 9 and 11. In trench 4 of the same site, it occurs between level 7 and 8. Chipped stones and flakes are common in this assemblage. In both cases the pottery is highly fragmented and thinner than any of the pottery above it. Although there was no clear decoration recovered to match those from the south, the fabric and forms are quite similar. This tradition has shown some affinities with the Rift valley PIW, especially in terms of the designs. Larger parts of Mnaida 1 and Mnaida 2 appear as single component sites. At Kitere, the tradition is overlaid by Proto-Swahili (PSW) tradition, to indicate a cultural gap that could have been caused by abandonment of the site for a period of time. Its occurrence in other sites is still not certain.

The distribution of EIW occupations along the Southern coast is witnessed on three sites: Pemba in Mikindani area; Mnangole and Kilwa Island. At Pemba and Mnangole the tradition forms the lowest pottery horizon. At Kilwa Island the tradition succeeded the PIW pottery tradition before being superseded by the TIW tradition, while at Mnangole there is a cultural gap after EIW period until the time of PSW.

Later Iron-working phases.

This section sequentially considers Triangular Incised Ware (TIW), Plain ware (PW) and Proto-Swahili Ware (PSW). They largely fall within the Chittick's (1966, 1974) 'Kitchen ware', category especially Period II and I.

The occurrence of the TIW tradition in the research area is apparent but not as common as other traditions. Its presence is witnessed as the lowest cultural horizon at Kabisela site in Mikindani. At Pemba, also in Mikindani, the tradition appears briefly above the EIW ware. It becomes more distinctive at Kilwa Island, previously identified as the lowest pottery tradition in Chittick's (1966, 1974) Period I. This work has demonstrated two more traditions below and predating the TIW. The population in TIW sites as well as their material remains is generally very low, as compared with the central coast of Tanzania. Imported exotic goods are very rare, suggesting the "pre-Islamic" or the early phase of TIW on that coast (Chami 1998).

The Plain ware tradition is also witnessed to exist above the TIW in the sequence. It appears distinctively among other pottery in the sequence by having extraordinarily thin walls, a rough surface and without decorations. The tradition occurs at higher frequencies in Mtwara than in Lindi region. Nakatumbatu ridge at Mikindani forms a type-site of the Plain ware tradition, with the highest record of such material. It nearly overlapped with the TIW at Kabisela site. This tradition is well known to have existed on the central coast of Tanzania and on the Island of Mafia between the 11th and the early 13th Century AD (Chami 1998; Chami 1999c).

The PSW is a new pottery tradition, herein defined as a distinctive chronological type that occurs above the Plain ware. The tradition is now confirmed to have existed in a wider area of the coast south of the Rufiji delta (Fig. 6.16). At Chikayowa site in Mtwara and Rushungi in Lindi, it forms the lowest pottery horizon. At Mnaida 3 and Kitere sites it is superimposed on the PIW Mnaida tradition, though in a disturbed condition in some of the trenches. At Mnangole site, the tradition is found above the EIW ware. At Kilwa Island, PSW materials occur just above those of the Plain ware. Above this tradition is the Swahili ware dominated by vessels with carinated forms, vestigial rims and commonly impressed beading around the neck, together with a high frequency of imported trade goods.

Relative dating

The MSA cultural tradition of the Southern coast has not yet been dated. The formal tools are clearly similar to those defined under Sangoan (Smola 1962, Isaac 1974). The Levalloisian flaking technique that provides tortoise-shaped cores and flakes and well-trimmed triangular points are characteristics of the Sangoan tradition common in the woodland south of the equator during the Middle Stone Age (Phillipson 2005). This tradition was widespread in the woodland by 300,000 BP and survived for a long time (ibid.). It is difficult at this stage to suggest the age of the southern coast Sangoan under discussion, but it may have survived in some areas of the woodland during and before the Holocene period.

Similar to this is the microlithic industry that is found below pottery horizons at both Lindi and Mtwara sites. In several literature, it has been related to 'wilton' microlithic industry of the late Stone Age in many parts of eastern and southern Africa. LSA occurrences are reported in many areas of the rift valley, south-western Kenya and north-eastern Tanzania. It is said to exhibit marked continuity between the pre-PIW hunter-gatherer phase and early settled cultivators (e.g. Ambrose 1984, Bower & Chadderdon 1986; Bower & Nelson 1978; Mehlman 1989).

In general, microlith industries were generally ubiquitous in eastern Africa, with general similarities and a confusing complexity of variations. No really convincing and meaningful classification of them has yet been proposed. The material used for making the tools is generally quartz of rather an informal character and occasionally chert. Although the available data are from widely scattered sites with quite different environments, the succession of post-MSA Sangoan stone industries in eastern Africa is now seen to have been broadly analogous to that revealed by more intensive research further to the south of Africa. These stone tool sequences continue with little or no modification during the PIW and in EIW periods (Huffman 2005; Sinclair et al. 1993; Holl, A. 1993; Soper 1971:8; Lane 2004).

The PIW correlates to the tradition of early settled communities of south-eastern Africa. The tradition is broadly contemporary with the late food-producing Pastoral-Neolithic communities of the Rift valley, with some shared designs and styles. The tradition predates the EIW traditions of the central and southern coasts of Tanzania. The "tillers of the soil" mentioned in the ancient Greco-Roman documents (Casson 1989) are here considered as responsible for the tradition. Taking into account Lacroix's (1998) arguments about Rhapta, and its location towards the southern coast, the tradition must have been well established on the Southern coast before the spread of the Phillipson's (1977) 'Chifumbaze' EIW complex. The adoption of the south-eastern Asian food crops on the south-eastern coast of Africa during the first centuries AD (Posnansky 1962; Oliver 1966) gives the upper limits of the tradition before being supplanted by the EIW culture.

The examination of the EIW ceramics from Southern coast of Tanzania shows a stronger affinity with the EIW found primarily in Malawi, especially southern tip of Lake Nyasa, eastern Zambia, and northern Zimbabwe. Chami (2006 : 119 - 128) identified the late phase pf EIW pottery tradition of the south coast, Mwangia, to be affiliated with the mkope / Gogomere tradition of the southern region of Africa, Malawi, Zambia & Zimbabwe. Machili site in Zambia comprises the occurrence of EIW variants dating to about 99 ±212 AD. False relief chevrons are common below the bevels found on the rims of the pottery but the number of bevels is not more than three. The thickening of the rim and the number of bevels on the rim causes it to resemble Limbo EIW tradition more than Kwale tradition. However, Nkope EIW variant at the southern end of Lake Nyasa has been reported to have these resemblances as well (Soper 1971: 25). Radiocarbon dates from Malawi, eastern Zambia and Zimbambwe have provided dates of the 4th century AD for the tradition (Phillipson 1975: 328)

The TIW tradition is widely known on the central coast of Tanzania (Chami 1994: 1998) but also further up the northern coast of East Africa, such as Manda (Chittick 1984) and Shanga (Hurton 1984). In southern Mozambique a similar pottery tradition is known in the Chibuene area (Sinclair 1982). In all of the TIW sites evidence of copper working or objects have been reported (Chami 1994: 44; Chittick 1974: 438-59; Sinclair 1982: 162). Other reported areas include Manda on the coast of Kenya (Chittick 1984; Abungu 1989: 174).

The tradition has been well discussed and dated, especially from the northern and central coasts of Tanzania (see Chami 1994) where two phases of the tradition Early or later, have been established. TIW dates range between 600AD and 1000 AD with the later phase associated with a high frequency of exotic trade goods Horton (1996) considers the later phase to be Islamic. The absence of any typically Islamic goods associated with the TIW pottery of the Southern coast indicates a pre-Islamic period for this tradition. Chittick's (1966:9) radiocarbon date of AD 630 ±110 from Kilwa Island corresponds to the early phase of this tradition (Chami 1994). It is now clear that a similar tradition extended as far as the southern coast of Tanzania and beyond (Sinclair 1982, Wright 1984 , Chami 2006). The TIW pottery from Kabisela is similar to that of Period Ia at Kilwa (Chittick 1966, Fig. 3a; 1974, Fig.95).

The Plain ware tradition partly correlates to the Kilwa Period Ib phase of the 'Kitchen ware' (Chittick 1974). However, in the latter, it was mixed with some TIW pottery that extends to Period Ia. The tradition shows little variation across the region. It forms a distinctive entity between 11th and 13th centuries on central coast of Tanzania (Chami 1998),

The Proto-Swahili ware (PSW) correlates to the period II in Chittick's classification of the Kilwa material (Chittick 1974, Fig.104) that spanned almost between the 12th and 14th centuries AD. One of the interesting forms is the vessel shape with double-curving profile from Mnang'ole and Rushungi. They were also recovered at Masakasa, similarly to those reported as period II from Sanje Kati (Chittick 1966, Fig. 6b; 1974, Fig. 104). Copper working was quite common during this tradition at most sites. A large number of crucibles used for copper working from Kilwa, Mnang'ole and Rushungi offer the best examples. One coin was found at Masakasa excavation in association with the PSW pottery (Fig. 6.17), similar to what have been reported from Kilwa, especially by Chittick (ibid.). Nevertheless there is no matching tradition on any of the central and northern coasts of the region. Paul Sinclair (pers. comm.) mentioned the occurrence of similar pottery on the northern coast of Mozambique.

Fig. 6.17: A Copper coin from Proto-Swahili horizon at Masakasa, Kilwa Island

6.4.3: Radio Carbon Dating

Charcoal samples were collected from different levels during excavations. A total of 20 charcoal samples were submitted to the Angström laboratory at Uppsala University (Sweden) for C-14 analysis. With the exception of the MSA and TIW levels, at least one date was obtained for each cultural type. Four of the samples contained a percentage of modern contamination, and two other samples were anonymously of a young age of the 20th century AD (**Ua 35919** and **Ua 35920**). The general results of these analyses are summarised below in Table 6.14.

Lab.No.	Sample	Trench site	Trench/Depth (cm)		^{14}C-age BP	Dates (uncalibr.)	Tradition
Ua 33159	M2	Mnaida 2	M2a	40 - 50	2195 +/- 35	245 +/- 35 BC	PIW
Ua 33160	M7	Nakatumbatu	M7	20 - 30	1070 +/- 35	AD 880 +/- 35	Plain ware
Ua 33161	M4 (1)	Pemba	M4	60 - 70	1560 +/- 35	AD 390 +/- 35	EIW
Ua 33162	M4 (2)	Pemba	M4	40 - 50	1480 +/- 35	AD 470 +/- 35	EIW
Ua 33163	M6 (1)	Chikayowa	M6	40 - 50	785 +/- 35	AD 1165 +/- 35	PSW
Ua 33164	M6 (2)	Chikayowa	M6	60 - 70	830 +/- 35	AD 1120 +/- 35	PSW
Ua 34100	R1	Rushungi	R1	40 - 50	825 +/- 40	AD 1125 +/- 40	PSW
Ua 34101	R2	Rushungi	R1	50 - 60	825 +/- 35	AD 1125 +/- 35	PSW
Ua 34104	R5	Rushungi	R4	40 - 50	390 +/- 40	AD 1560 +/- 40	Swahili
Ua 34102	R3	Rushungi	R2	20 - 30	112.7 +/- 0.5pM		
Ua 34103	R4	Rushungi	R3	30 - 40	114.3 +/- 0.5pM		
Ua 35915	Masak 3	Kilwa Island	T3	100 -110	2690 +/- 35	740 +/- 35 BC	PIW
Ua 35916	Masak 4	Kilwa Island	T4	80 - 90	1630 +/- 35	AD 320 +/- 35	EIW
Ua 35918	Mnan 1-9	Mnangole	T9	50 - 60	4520 +/- 40	2570 +/-40 BC	LSA
Ua 35921	Kite 1 (1)	Kitere	T1	40 - 50	585 +/- 35	AD 1365 +/- 35	PSW
Ua 35922	Kite 1 (2)	Kitere	T1	50 - 60	825 +/- 45	AD 1125 +/- 45	PSW
Ua 35914	Masak 1	Kilwa Island	T2	60 - 70	146.6 +/- 0.6pM		
Ua 35917	Mnan 4	Mnangole	T5E	70 - 80	101.3 +/- 0.5pM		
Ua 35919	Mnaid 2-6	Mnaida 2	T1	40 - 50	55 +/- 35		
Ua 35920	Mnaid 2-6	Mnaida 2	T6	50 - 60	60 +/- 35		

Table 6.14: Carbon-14 Dates from Southern coast and their Cultural implications

Lab no.	Sample	Site Trench	Trench	Depth	14C age BP	Dates	Tradition
GX 0398	KK 1	Kilwa Island			1825 +/-110	AD 125+/-110	EIW
SR 77	KK 3	Kilwa Island			1320 +/-110	AD 630+/-110	TIW
SR 78	KK 4	Kilwa Island			1020 +/-100	AD 930+/-100	Plain wares
SR 76	KK 2	Kilwa Island			890 +/-100	AD 1060+/-100	PSW
N-256	KK 5	Kilwa Island			815 +/-110	AD 1135+/-110	PSW

Table 6.15: Previous C-14 Dates from Kilwa by Chittick (1966)

The obtained date of 4520 +/-40 BP for the late Stone Age culture at Mnang'ole does not enable the time range of the tradition to be established. It does not seem to represent the early or even the late phase of that tradition. Similar case is that of the PIW pottery of the region. The earliest date is that from Masakasa site at Kilwa Island. The charcoal sample obtained within the pottery assemblage at a depth of 107cm in trench 3 of Masakasa (Ua 35915) provided a date of 2690 +/-35 BP. This date implies a settlement of the 8th century BC. The epicentre of this site could not be established, suggesting the possibility of an even older date in the future that could correspond to the beginning of better climatic condition soon after about 2000 BC drought (Lonnie et al. 2002). The date from Mnaida hill at Mikindani had suggested an age of 3rd century BC for the tradition for the lowest pottery horizons, which are from the lowest pottery level of 40-50 cm of trench M2a at Mnaida 2 (sample Ua 33159 in Table 6.14). Together, the two dates offer a wide range of time for the generalised pre-EIW settled tradition (PIW). It is reasonably safe to argue here that these dates are neither 'too early" nor do they set limits for the temporal distribution of the PIW tradition. They closely correspond to the early non-archaeological accounts of the coast (Lacroix 1998, Casson 1989, Davidson 1964, Baxter 1944).

Three dates for the EIW tradition were obtained in this work. Two samples from levels 7 and 5 in trench M4 at Pemba site (Ua 33161 and Ua 33162) provided an age of the 4th and 5th centuries AD for the tradition (Table 6.14). The excavation could not recover charcoal from the 8th level that bears pottery as well. Logically, the retrieved dates do not offer the beginning of this tradition. From Masakasa at Kilwa, another date of the early 4th century AD was obtained from sample Ua 35916 (Table 6.14). These EIW dates are nearly the same as those of early Nkope tradition especially Gokomere facies ranging between the 3rd and 5th centuries AD (Chapter 7, Table 7.1). Chittick's (1966) excavations obtained the date of the AD 125 +/-110 (GX 0398) at Kilwa (Table 6.15) but this date was rejected for what was argued as being too early.

The TIW horizons were not conclusively dated in this work. However, from the Kilwa Island a date of AD 630 +/-110 (SR 77) was obtained (Table 6.15), but once again rejected for been "much too early" for the lower 'kitchen ware' (Chittick 1966:9). The lower 'kitchen ware' is now confirmed to be the TIW pre-Islamic phase. The early TIW phase is also argued to date from the 4th to the end of the 7th century, a period associated with Sassanid trade goods (Chami 1994:94). At Kabisela site, the TIW assemblage is found in association with Sassanid pottery (Chapter 5, Fig. 5.32, C). In this regard, the age of the Southern coast TIW phase is the same as that of the central coast of East Africa (Chami 1998) and southern Africa (Sinclair at al. 1993).

There is very little discussion in the literature about Plain ware tradition (Chami 1998; 1999c). A well-defined period for this tradition is between 10th and 13th centuries AD (Chami 1998). Nakatumbatu site is a single-component site fully occupied by people with the PW tradition. A Charcoal sample from an undisturbed sequence yielded an age of AD 880 +/-35. Similar pottery was found in level 6 of Masakasa trench 3 and 3E, but also level 4 of trench 2 of the same site. The tradition occurs just above the TIW pottery. The previous date of AD 930 +/-100 from Kilwa (Table 6.15) closely corresponds to the Nakatumbatu date. Although the Kilwa date could in the same way represent the late/Islamic phase of TIW, there is nothing at all at Nakatumbatu to correlate to a TIW tradition, but rather to the Plain Ware, which seems to have flourished on the southern coast between the 9th and 11th centuries, before being succeeded by the PSW in the late 11th century.

The PSW pottery has been dated more thoroughly than the rest of the pottery horizons. Six new dates were obtained from Chikayowa, Rushungi, and Kitere sites (in Table 6.14). At Kilwa, the tradition had previously yielded two accepted dates of the late 11th and early 12th centuries (Table 6.15). Except for the mid-14th c date from Kitere, the rest of the dates clustered in the 1st half of the 12th century. A copper coin was found in this assemblage at Kilwa Island during this work (Fig. 6.15). Neville Chittick (1966) attributed this cultural phase of period II to the reign of the first dynasty. Based on this consistency of dates and patterns, the coin is believed to belong to the same period of the first dynasty.

Based on these radiometric dates, the chronology of the Southern coast has been reinstated to indicate continued human activities on this coast since the Stone Age as summarized in Table 6.16 below. It closely matches the sequence established by the cultural materials (Fig.6.14) and other known archaeological and linguistic data in eastern and southern Africa. The research hypotheses have clearly been verified accordingly.

CULTURAL PHASES AND C-14 DATES			
Swahili wares	AD 1560 +/-40[14]			
PSW	AD 1120 +/-35[8]	AD 1165 +/-35[9]	AD 1125 +/- 40[10]	AD 1125 +/-35[11]
			AD 1365 +/- 35[12]	AD 1125 +/-45[13]
Plain Wares	AD 880 +/-35[7]	AD 930 +/-100		
TIW	AD 630 +/-110			
EIW	AD 320 +/-35[4]	AD 390 +/-35[5]	AD 470 +/-35[6]	
PIW	BC 740 +/-35[2]	BC 245 +/-35[3]		
LSA	BC 2570 +/-40[1]			

Table 6.16: Reconstructed Cultural Phases in Chronometric Sequence.

<u>Notes</u>: *1 Mnang'ole; 2, 4 Kilwa Island; 3 Mnaida-2; 5-6 Pemba (Mikindani); 7 Nakatumbatu; 8-9 Chikayowa; 10-11, 14 Rushungi; 12-13 Kitere.*

Chapter **7**

DISCUSSION AND CONCLUSION

This work has endeavored to establish the Holocene archaeological sequence of the southern coast of Tanzania. In order to answer the specific problems of the research, the work aimed at several targets that will be reviewed in this chapter in order to measure the success.

The work aimed at a thorough examination of the coast to prove the presence of stone cultures and even settled traditions prior to the spread of EIW technologies. colleague and I (Chami & Kwekason 2003) had reported that sites in this region had what was considered as pre-Ironworking (PIW) pottery. These potteries were thought to be related to stone-working sedentary settlements. Cultural contacts or relationships between the southern coast and the Rift Valley in the interior, either through interactions or exchange of ideas and materials, were suggested due to the presence of some similar cultural elements in their pottery. As argued by Huffman (1980), pottery style, as part of culture, is learned and possessed within groups of people, and the correlation between design style and specific groups is well known. Rice (1987: 252) suggested that similarity of design elements between groups was proportional to the direction and intensity of social interactions between members of these groups. Tilley (1991 : 1.5) argued that it is the similarity and their subsequent relationship which matters and not mere elements or meaning. Tilley (1991: 105), argued that specific identity of the designs – the content, is not so important because meaning arises from the relationship between designs, not from any specific design in isolation. Arguing along these lines, this work aimed at describing and establishing the temporal and spatial relationships of the early settled traditions of the coast.

Secondly, it aimed at examining the cultural sequences at different sites on the coast to establish their diachronic and synchronic patterning, so as to verify the stratigraphic position of the EIW pottery, which has been argued to be the earliest pottery on southern Tanzania (Phillipson, 1977; 2005). Therefore, there was a need to provide widespread evidences in order to justify our previous argument (Chami & Kwekason 2003) that the southern coast had stone-age settlements with pottery making skills, and hence settled communities, prior to the spread of the EIW pottery tradition and other associated cultural package.

The work also wished to testify the alleged "tillers of soils" on the southern coast during the time of the Periplus (Casson 1989). The argument was that these were not Bantu speekers who had settled on the land before the EIW communities, and alleged to have adopted and developed Southeast Asian agriculture and its spread into the interior (Oliver 1966). Wrigley (1960: 201), who examined Guthrie's (1962) linguistic results on the early spread of the Bantu speakers along the woodland south of the equatorial forests, suggested that the EIW agricultural revolution of the early centuries AD was a later event over already settled land.

> "… I see these people, not as agriculturalists spreading over a virtually empty land, but as a dominant minority specialised in hunting with spears, constantly attracting new adherents by their fabulous prestige as suppliers of meat, constantly throwing off new bands of migratory adventurers, until the whole southern sub-continent was iron-using and Bantu speaking" (Wrigley, 1960:201).

This work aimed at defining the culture of these earlier settlers and compare their pottery tradition against the later and more famous EIW pottery tradition.

Lastly, the work wished to establish the cultural relationship between the southern coast and further interior, or the immediate northern coast. For sometimes, schoolars thought the coast of eastern Africa had existed independent of the mainland. Hence this work intended to find the affiliations of the coastal population to the traditions of the southern woodland that extends almost unbroken across southern Tanzania to the Atlantic coast in the west. Other than that, there was a wish to examine any alleged early Cushitic speakers existence in this area.

The results of these examinations and evaluations led to the establishment of the Holocene archaeological sequence of the southern coast as discussed in little the following sections. It should be reiterated here that this work contributes very little knowlldge to the early and middle parts of the Holocene. Albeit the thorough search of the established data universe as discussed in this thesis, the work could not reveal the extent of the Stone Age cultures during the mentioned periods. It is therefore fair to argue that the wetter conditions that are known to have prevailed and dominated larger part of the tropical Africa between 11,700 and 4,000 BP (Lonnie et al. 2002; Maley 1993), which caused sedentary life on larger parts of Africa is not testified in this region. It is after between 4,000 and 3,000 BP when modern sendentary conditions become established on the southern coast, when sedentary life started to appear in most of the woodlands and coastal areas.

This work has demonstrated the presence of Stone Age cultures on several sites in the research area, some of which could have flourished long before the Holocene. It is necessary to have them discussed here as part of the finds since the MSA and LSA cultures are known to have co-existed for an unknown period of time in large parts of the sub-Saharan region until very late (McBrearty & Brooks, 2000).

It is not the first time they are reported from the East African Coast. They were reported from Mombasa (Omi 1982) and Bagamoyo (Chami 1996) as indicated earlier from Smola (1962) and Isaac (1974) for Kilwa Lindi region.

Distributions of MSA- derived stone tools were seen in most parts of the coastal woodlands of the research area. Although excavations and radiometric dating could not demonstrate their diachronic pattern, the spatial distribution of the tool types was observed at Mpara area of Kilwa mainland, north-eastern corner of the Kilwa Island, woodland areas along the Mbwemkuru River down to the shore, the woodland area of Mchinga, and also cleared areas of Mikindani. Triangular points and scrapers on clearly Levalloisian flakes are vivid examples proving the presence of this tradition, although in few amounts and in scattered condition.

The Upper Mbwemkuru seems to form the core for this tradition, which is not surprising since a reasonable amount of unpublished MSA implements from the southeastern regions of Tanzania are preserved in the National Museum of Tanzania, Dar es Salaam. These include MSA implements collected near Kipili area, 80 km from Lindi town along the Masasi road (Fig. 5.02). A similar assemblage was collected from Kimbilimu hill on the road from Milola to Nangaru in Lindi district. Among the tools are bifacial tools, Levalloisian cores, points and scrapers of typical MSA tradition, very similar to what was described by Smolla (1962) and Isaac (1974). More tools were collected from Mandawa village; Chikotwa area and Kiliamanda hill (Fig. 5.02). Others are from Makonde beds in Mikadi area of Newala district. The new collection during this research offers a clearer picture that MSA population that paved the way for the Holocene LSA communities occupied the region. Mehlman (1979), who described MSA culture from Eyasi and Serengeti sites, seems to document the transition to the LSA, the beginning of which was dated very roughly to about 20,000BP.

It should also be noted here that, while I was writting this work, Chami (2009a) excavated several caves in Zanzibar and Mafia where he collected stone tools analysed by several schoolars, including Knutsson (2007) and Kessy (2009). The date of the earliest horizon with MSA tools are between 20,000 and 30,000 years ago. It is argued by Chami (2009b) that MSA population could have spread to all Islands as far as Madagascar. This suggest that MSA population could also have sailed and exploited various marine resources.

Between 7,000 and 4,800 BP a ceramic microlithic MSA culture is documented in several parts of the area.

The lowest occupation level during these excavations is a ceramic LSA-related tradition that continued in some areas during the ceramic tradition. This could be part of the microlithic industries virtually ubiquitous in southern Africa, conceivably linked with the widespread adaptation of bow and arrow hunting techniques (Mitchell 2002). Late Stone Age horizons were clearly observed in Mchinga-Mnang'ole areas, Kilwa and also Kitere areas. In the Upper Mbwemkuru valley, a number of flaked pieces were seen on the slopes of eroding gullies, some of which are presented in this work. A number of crudely struck flakes of quartz form the majority of the assemblages of small, fairly delicate flakes of siliceous material.

Formal tools are generally rare in most of the sites but cores and core fragments, especially bipolar and single platform types, are part of the Masakasa assemblage. Some of the cores could only have been properly flaked by placing them on an anvil due to its relatively small size. Backed or trimmed tools are even rarer although large number of flakes, blades and other waste products can be seen. What was observed here is similar to the reported Serengeti situation (Mehlman 1979), where backed implements were absent in most sites located in the tall grasses. Some of these lithic materials continued to appear in the higher pottery-bearing horizons. The maker of these artefacts seems to have only depended solely on the local source of the raw material, which is composed of relatively small quartz and chert pebbles. There is a reasonable doubt here in assigning this tradition to either of the LSA variants described in eastern Africa due to the too small amount of formal tools contained in the collection.

It should also be noted that finds from Zanzibar indicate Stone Age horizon with absense of geometric tools (Kessy 2009). However, different from these of my research area, the caves in Zanzibar have worked bone tools including harpoon and decorated bone typical of LSA with ceramic in the Nile Valley region (Chami 2009b). Probably the absence of these in my reseach area is due to poor preserved conditions, but this does not explain absence of geometric tools.

The above two stone Age -working cultures remain poorly dated. Dating of such cultures has never been a simple task due to lack of proper dating materials. Charcoals were rare in the lithic horizons. Only one date was obtained that could represent this period. From Mnangole site, sample Ua 35918 provided an age of 4,520 ± 20 PB. This is one of the commendable dates, but it does not provide a better understanding of the timeframe of this tradition, though it matches the upper limit of the Mehlman's (1979; 1989) microlithic industry. In Zanzibar the PIW culture dates to the 6000 BP (Chami 2009b) similar to the beginning of such culture in the Nile Valley or the Rift Valley.

Similarly it is difficult to suggest the socio-economic life of these Stone Age communities based on the recovered evidence. They must have formed bands of hunters and fishers along the coast and in the woodland, as suggested by hundreds of small, delicate flakes and broken cores recovered at Masakasa, Kitere and Mnangole excavations. Erosion and rains wash-away of the covering soils were seen re-concentrating formal tools on the surface. Again it should be noted that in Zanzibar and other Islands of the Indian Ocean evidence of animal domestication and trade is present (Chami 2009b). Fish bones and shells both marine and land were recovered in good amount (Chami 2009b).

On the other hand, the low concentration of tools and other cultural materials or artefacts could suggest low population of the Stone Age communities that was scattered across a wide area. The evidences for their activities along the shore had no chance of leaving behind visible features or objects because of the activeness of the beach processes. This situation leaves us with little understanding about the culture of the people under discussion. The continuation of stone working even during the Ironworking period could suggest continued adaptation and assimilation by the same population. Population could have been moving from one area to another, adapting new techinques of life rather than displacement by an entirely new ethinic population such as the allenged cushitic speakers (Ehret 1998). Although metallurgy was embraced, stone tool technologies and foraging persisted in several areas of East Africa (see also Kusimba & Kusimba 2005).

It has been argued elsewhere that the southern coast of Tanzania has provided evidence of pre-EIW settled traditions with pottery making technology (Chami & Kwekason 2003; Kwekason 2007). These are, in other words, Stone-working settled traditions collectively defined in this thesis as PIW culture. The potteries of these traditions are described as Pre-Ironworking ware (PIW).

The clearest evidence of a PIW of the southern coast is the Mnaida tradition that spread from the littoral at Mnaida hill (Mikindani) to the hinterland in the Kitere area. It is widely accepted that the invention of pottery more-or-less coincided with the beginning of more lasting or sedentary settlements (Sutton 1974, Phillipson 2005). Broken ceramic vessels (potsherds) are among the most common archaeological finds that have been used in these circumstances to indicate settlements. Their distributions and frequencies are said to suggest the nature of the settlement in question. Their shape, style, and form have provided the foundations for thousands of archaeological analyses (Orton et al. 1993; Rice 1987).

Archaelogists have considered pottery as synomym for agriculture (Seddom 1968). However, there are a few contrasting cases where, for instance, in Japan, pottery was made by hunter-gatherers before 10,000 BC (Akazawa & Aikens 1986) and in Mexico where cultivation began before the first pottery appeared (Smith, B.D. 1999). In central Sahara (now desert), harpoon fishers are known to have made pottery before farming and domestication (Phillipson 2005). In lake Turkana (Barthelme 1985) and also in Kiwangwa - Bagamoyo (Chami 1996) on the coast, pottery has been dated to more than 10,000 BC.

Repeated combinations of different motifs in different positions are said to produce a kind of group identity, indicating a settled community (Huffman, 2005). The skills of manufacturing pottery and the usefulness of such fragile vessels have been closely related to a reasonably settled life (Sutton 1974: 530). Therefore, scattering of pottery were used to suggest the beginning of a settled life on the southern coast, roughly during the last two millennium BC, representing a cultural change or invention in stone-working population or implying the alleged Cushitic or Bantu speakers migration.

Mnaida hill at Mikindani provides the best site for pottery of this tradition, especially Mnaida 1 and Mnaida 2, and hence Mnaida tradition. The radiocarbon dating of a horizon with this pottery type has provided an age of the 3rd century BC (245 $\pm$ 35) (see Table 6.14). Their potteries are quite distinctive, with thin walls and cuneiform stamping decorations. Closely packed cuneiform stamps and rocker stamping form parallel straight, looped or chevron lines in a regular manner from the lip. Most common shapes are globular vessels and open bowls with a wide orifice and shallow depth (Fig. 5.33-44; 6.07:2a &b). Similar pottery is found at Kitere hill (Fig. 5.41).

At Kilwa, below an EIW horizon was similarly thin-walled pottery as opposed to the thick EIW pottery above it. In trench 3 of Masakasa site it appears between levels 9 and 11. In trench 4 of the same site it occurs between levels 7 and 8. In both cases, the pottery was more fragmented than any of the pottery above it. Although no decoration that was recovered to match that from Mtwara region, charcoal sample from the lower pottery-bearing horizon was dated to 740±35 BC. This pottery was certainly superseded by the EIW in the same section of undisturbed sequence. Its occurrence on other sites is still not certain.

The 8th century BC date is just a scientific proof to add to what had been suggested by other scholars. For instance, Davidson (1964: 173-5) suggested that settlements had already been established on the coast of Tanzania as early as the middle of the last millennium BC. Baxter (1944) argued for an established how trade between the settled East African coast and other northern states by the end of the 7th century BC. Such early dates were rejected as dubious due to the absence of supporting archaeological evidence. From the interior of the southern woodland the association of pottery and lithic has been dated to as early as 600 BC (Mazel 1992). This is the same almost unbroken woodland that extends to the research area on the southern coast.

It should be noted here that it is not archaeology alone which provides evidence for settled life on the Swahili coast before the EIW period. Chami (2006, 2009b) has done a detailed discussion of ancient records dating back of the Pharaonic Egypt. Records indicate that the lands of Punt, dating back to 3000BC, and being mentioned as late as 700BC, extended to the region of my study (Bolliger 2005). Panchea, which preceeded Azania, well detailed by Iambulus or Euhemerus (see Chami 2006) reported of settled population, practising both pastoral and agricultural life of about 200 - 400 BC.

Chipped stones, cores and flakes are common in these assemblages, not only at Kilwa but also in Mnangole and Mtwara sites. Researchers on coastal archaeology (e.g. Chami 1996 - 2009b), in the hinterland (e.g. Thorp 1992) but also more frequently in the Rift valley region (e.g. Ambrose 1984, Bower & Chadderdon 1986; Bower & Nelson 1978; Mehlman 1989), have reported a marked continuity between LSA and the EIW pottery traditions. The only noticeable change in the lithic artefacts is their decreasing frequency up the sequence than a stylistic variation.

This work could not come up with any evidence suggesting the possession of iron technology by the PIW communities. These early settlers might have used iron objects without smelting knowledge, as suggested by other sources (see Kwekason & Chami 2003). *Periplus of the Erythrean Sea*, for instance, reported about importation of iron objects such as lances, hatchets, swords and awls from Arabia. These could have been nicely finished and prestigious products of higher status than the local products as it is done today. The finished iron products produced through casting could have received very little competition against local products given the nature of forging that was typical in most African ironworking (Miller & van de Merwe 1994; Schmidt 1997).

The distribution of the EIW culture along the southern coast of Tanzania seems to be in restricted areas. From Kilwa to Ruvuma basin, only a few sites were discovered. Their temporal distribution is of the same nature as those of other known similar sites in southern Africa, but their cultural materials do not show clear continuity with the preceding culture. Most of the earlier settlements had been abandoned before this phase, perhaps due to soil degradation or drought. Comb stamping that forms a band of parallel oblique lines, broad grooves and rows of single impressions is a common decoration on their pottery, and is usually around the rim but in a few instances along the shoulder (Fig.7.2). Occasionally there is a series of pendants around the shoulder of necked pots. Fine incisions appear to be absent or very rare, although their appearance increases as one moves north of the research area. False relief chevrons are said to have been quite common in the early stage of Kapwirimbwe and Chondwe facies of Nkope EIW tradition, as well as in Kwale tradition, which are all variants of EIW cultural complex (Soper 1971). While at Kilwa, these false relief chevrons are combined with clear bevelling without a band around the rim, further to the south they appear with comb stamping around the rim.

Variations among the sites have been noticed. In Mtwara sites the bands are entirely filled by broad grooving, oblique or interlocking blocks of parallel lines. The observed series of triangular pendants are filled by horizontal grooves. Bordering is always by stamping or a line of impressions. Further north in Mchinga-Mnangole areas, filling by comb stamping become more common. The interlocking blocks are lacking and the pendants are filled by vertical grooves. The designs and borderlines are almost of the same kind. The lowest EIW materials showed some false relief chevrons that appear below bevelling, with the number of bevels limited to a maximum of three. The Kilwa EIW has a startling resemblance to Kwale tradition in certain specific traits, such as bevelling and up-turned rim bowls. It is apparent that the influence of Kwale tradition intensifies northwards toward central coast, where it becomes typical Kwale.

In general, on the southern coast, the EIW pottery with false relief chevrons below the bevels was not common. The number of bevels is frequently not more than three. The thickening of the rim and the number of bevels on the rim makes it resemble Limbo more than Kwale traditions. In relating this tradition to those of the central coast of Tanzania, the radiocarbon dates and the nature of rim thickening suggest a Mwangia phase of EIW pottery. The Style and design of this pottery does not show most of the Mwangia features like shoulder fluting and borderless grooving. The elaborated rim platform for carrying a band of decorations, common in this type of pottery is more typical of the interior woodland sites (Fig. 7.3) than on the northern coast of Tanzania. Occasionally there are certain affinities to Limbo ware which is the earliest EIW phase on the central coast of Tanzania (Fig. 7.1) than of Kwale that succeeded it, although the boldness of the grooving or channelling and radiometric dating does not correspond to any of the Limbo EIW tradition.

However, Nkope variant of the southern end of Lake Nyasa (Fig.7.4) has been reported to have these resemblances (Soper 1971: 25). Gokomere EIW features are said to dominate in the EIW pottery of northern Mozambique (Sinclair 1987; Sinclair et al. 1993). Evidence of the Kwale tradition seems to be quite rare not only on the southern coast Tanzania (this work) but also towards northern coast of Mozambique (Sinclair et al. 1993). The radiocarbon dates from the sites of the southern coastal Tanzania (Table 7.1, shaded) strongly correlate to those from Malawi and eastern Zambia of AD 320, AD 320 respectively, and also to Zimbabwe of AD 360 (Phillipson.1975: 328).

REGION	SITES	^{14}C DATES		
EAST AFRICAN COAST	Kwale/Kenya	AD 270 +/-115	AD 260 +/-115	
	Kwale Isl./Tanzania	AD 240 +/-60	AD 425 +/- 55	
	Mafia Isl./Tanzania	AD 240 +/-60		
	Kibiti/Tanzania	AD 280 +/-60	AD 315 +/-70	AD 315 +/-70
	Limbo/Tanzania	AD 233 +/-60	**BC** 167 +/-90	**BC** 93 +/-90
	Kivinja/Tanzania	AD 431 +/-70	AD 598 +/-70	
	Kilwa Isl./Tanzania	**AD 320 +/- 35**		
	Mnang'ole/Tanzania	**AD 320 +/- 35**		
	Mikindani/Tanzania	**AD 390 +/- 35**	**AD 470 +/- 35**	
SOUTH-EASTERN AFRICA	Machili/Zambia	AD 20 +/-100	AD 96 +/-212	AD 489 +/-82
	Silver Leaves/SA	AD 270 +/-60	AD 250 +/-40	AD 330 +/- 50
	Matola/Mozambique	AD 230 +/-110	AD 70 +/-50	AD 430 +/- 80
	Maputo/Mozamb.	AD 174 +/- 85	AD 360 +/-75	AD 615 +/-100
	Ziwa/Zimbwabwe	AD 300 +/-100		
	Ngwenya/ ?	AD 400 +/- 30	AD 400 +/-60	AD 415 +/- 30

Table 7.1: Comparison of EIW dates from sites of eastern and southern Africa (modified after Chami, 2006:120).

A number of sites show pottery of the EIW associated with the Late Stone Age tools with little or no modification (Huffman 2005; Sinclair et al. 1993; Holl, A. 1993; Soper 1971:8; Lane 2004). In the EIW communities grouped together as Chifumbaze (Phillipson 1976; 1993; 2005), a broad distinction exists between different sets of assemblages. Whether they form part of the same tradition or not, is still debatable (also Collett 1988). Nevertheless, the ceramic designs among many of the Chifumbaze (EIW) group are said to have widely shared features on a scale far greater than any conceivable political or socio-linguistic entity (Huffman 1980; 1989). Both chronological and typological evidences are still insufficient to support or prove the direction of the spread of these EIW groups and the interrelationships among the various geographical variants as argued by Phillipson, Huffman, etc (ibid.).

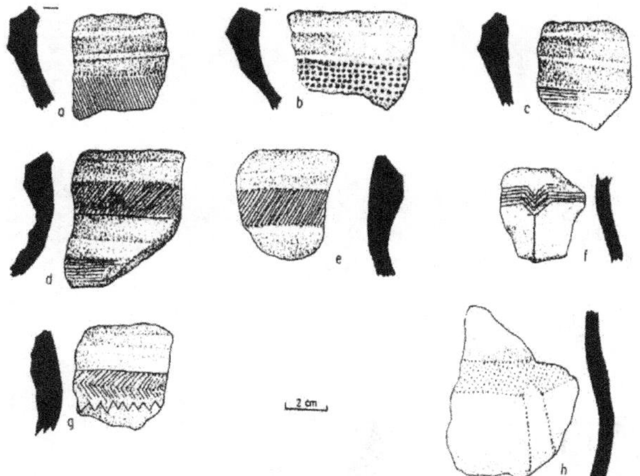

Fig. 7.1 Limbo EIW from central coast of Tanzania (after Chami 2006:120)

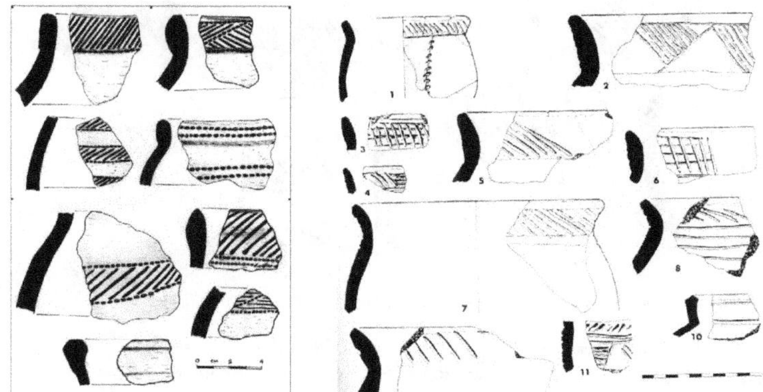

Fig. 7.2: Early Ironworking pottery from Mikindani area

Fig. 7.3: Mzonjani EIW pottery (in Mitchell, 2002:265)

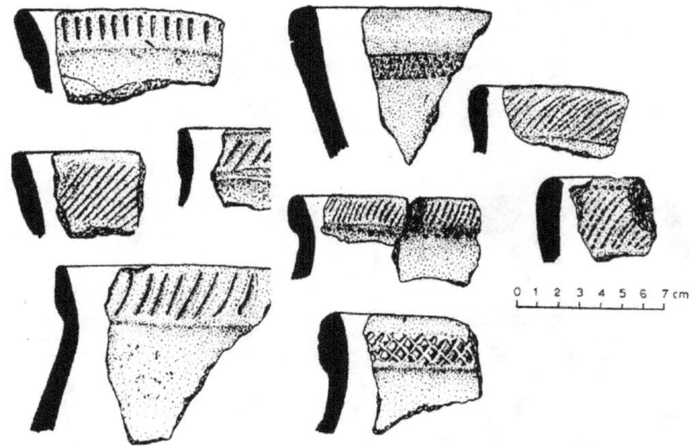

Fig. 7.4: Some of the early Nkope/Gokomere pottery tradition (after Phillipson 1976)

It has been long argued that EIW evidences from the southern tip of lake Nyasa suggested either a separate influence from the southern interior other than the Kwale from the northern coast (Soper1971: 21). Research on southern Tanzania and northern Mozambique was expected to provide evidence of links along the region to east of Lake Nyasa. This work has provided not only the links between the interior and coastal EIW, but also between the EIW and PIW Bantu communities. The EIW pottery of the southern coast does not seem to have resulted from the north-south spread of Kwale as previously assumed. The assumed stronger interior-coast trend of influence along the southern woodland (Soper's 1971: 25) is what is reflected by results of this work. The suggested domination of Gokomere stylistic features in the pottery of northern Mozambique without or with very little evidence of Kwale tradition is now truly a broader phenomenon of the southern Coast. It is the Nkope tradition which extends to the coast rather than kwale from the north. Chami's (2006) three phases of EIW tradition is not well demostrated in this area of my research. However, the appearance of Kwale phase pottery at Kilwa suggests that further research is required to better understand Chami's there phases of EIW culture.

The occurrence of the TIW tradition on the southern coast of Tanzania is apparent, but similarly rare as it is for the EIW tradition. At Kabisela site in Mikindani, the tradition appears as the lowest pottery horizon. Close to this location is the Pemba site, where the tradition appears briefly above EIW ware. The TIW pottery becomes more distinct at Kilwa Island, previously thought to form the lowest pottery horizon in Chittick's excavations (1966, 1974). The population of TIW sites, as well as related material, is generally very low compared with the central coast of Tanzania. Imported exotic goods are very rare on the southern coastal until after the PSW phase, though a few Indian glass beads, graffito wares and very thin glass ware forms interesting evidence of imports, both at Kilwa and Mikindani areas during the PSW period (Chapter Five).

The TIW tradition is widely known on the central coast of Tanzania (Chami 1994: 1998 Juma 2005), and further up the northern coast of East Africa, such as Manda (Chittick 1984) and Shanga (Hurton 1984). Chami (1994, 1998) in support by Juma (2005) , established 2 phases of TIW tradition on the central coast early TIW - 600 - 700 AD and later 700 - 900 AD. The early phase has yet to be established on the Kenya coast. The date in my research area recognises its early phase. In southern Mozambique, a similar pottery tradition is known from Chibuene area (Sinclair 1982) and in Comores (Wright 1984, Chami 2009a). In all of the TIW sites, evidence of copper working or objects has been reported (Chami 1994: 44; Chittick 1974: 438-59; 1984; Sinclair 1982: 162; Abungu 1989: 174). In this work most of the evidence that could be related to copper working like those refractory crucibles appears much higher up the sequence.

TIW tradition is also known to date between 600AD and 900 AD with its late phase assemblage associated with a lot of exotic trade goods from the Islamic world, hence 'Islamic phase' (Chami 1994 Horton 1996a). Chittick's radiocarbon date of AD 630 ±110 from Kilwa Island (1966:9) that was considered too early for the pottery tradition on the coast of East Africa seems to correspond to the early phase of this tradition. Their occurrence on the southern coast makes a link between eastern and southern Africa. From the late phase of TIW, ceramic technology largely changed under the influence of expanded importation of ceramic goods and techniques.

Chami (1998) introduced and dissussed Plain Ware tradition in detail. The Plain Ware tradition witnessed to exist above the TIW in the sequence. It appears distinctively from other pottery in the sequence by having extraordinarily thin walls and rough surface. The tradition occurs at higher frequencies in Mtwara areas than in Lindi region. Actually Chami (1988) argued that this tradition flourished in the region South of Bagamoyo - Zanzibar (see also Chami 2009b)

The Nakatumbatu ridge at Mikindani forms a type-site of the Plain ware tradition, with the highest record of such pottery. An extensive village that stretched for about a kilometre occupied the ridge, leaving behind a high concentration of local ceramic fragments, without a mixture of other cultural elements. The site seems to have been abandoned for a length of time until quite recently. A few PW families could have lived at Kabisela site, with very similar material in form and preservation condition. A charcoal sample associated with the lower richest pottery level at Nakatumbatu provided a date of AD 880 ±35. This is currently the earliest known age for such a tradition on the entire coast of East Africa since Chami (1998, 2009b) dated the tradition to 900 - 1200 AD.

This date has a starting similarily to one of the date from Kilwa Island (Table 6.15), of AD 930 ±100 for the 'Kitchen ware', which is now confirmed to have consisted of three distinctive cultural phases include the Plain Ware tradition. The Plain ware tradition is well known to have existed on the central coast of Tanzania and the Island of Mafia between the 900 and the early 12th Centuries AD (Chami 1998; Chami 1999c). In these circumstances, the date from Nakatumbatu makes the Plain Ware tradition of the southern coast the oldest known in East Africa, succeeding the TIW in most sites.

Plain Ware may have existed elsewhere on the northern coast, and even in southern Africa. Chami (pers.com) suggests that better research is required for the Kenya and Tanzania northern coast to deny its existence. In Chami (2009b) a report of similar pottery in Zimbabwe appears.

Proto-Swahili Ware is a new ceramic tradition defined as a distinctive tradition in this work it dates between the 12th and 14th centuries AD. Some of the ceramics that belong to this group were categorised as Period Ib and II of the early 'Kitchen ware' (Chittick 1974). The tradition is now confirmed to have existed in a wider area of the coast south of Rufiji delta of East Africa. A good number of graphic presentations of the pottery have been provided in Chapter Five of this book. Up to this stage, no similar pottery that has been reported from either the central or the northern coast of the region, although Paul Sinclair (Press. Comm.) mentioned that similar pottery designs exists on the northern coast of Mozambique, possibly contemporary with the one discussed here.

Interesting features in this tradition include the emphasis on festooned or arched bands around the rim-shoulder area. Shell impressions, hatching, and stamping or even roulettes forming similar patterns are quite common. On hemispherical and open bowl forms, the band is mainly from the lip in the form of hanging festoons or loops. On necked and globular forms, the patterns are mainly above the shoulder as arched or horizontal bands usually with indented lips. Some of the necked vessels have a double-curved profile, an unknown shape type in most described traditions.

Radiocarbon dating of the PSW tradition provided consistent results. Most of the dates are clustered in the early 12th century (Table 6.16). One date of a much higher level falls in the early 14th Century. Local variations are quite obvious as discussed in Chapter Six. The upper levels of this tradition overlap with the known Swahili ware with vestigial rim, neck or shoulder punctuates and carinated forms (see Chami 1998). Copper working was quite common during this tradition at most sites. A great number of crucibles used for copper working from the Kilwa, Mnang'ole and Rushungi sites offer the best examples. This type of vessels was hardly found in the lower sequences and completely missing during the EIW. Although assumed here that they were mainly used for copper working, the sources for raw material as well as smelting techniques are yet to be identified. The extent of this work could not find any type of furnace during the entire fieldwork, although high frequencies of iron slag and used crucible fragments were observed. It is only one coin that was found at Masakasa excavation in association with the PSW pottery (Fig.6.17) that demonstrates local copper working on the coast.

It is becoming more apparent that copper working was common during the Proto-Swahili period for production of small domestic items. Crucibles for melting copper have higher frequencies at the beginning of the second millennium AD than before. The beginning of this tradition is not clear but it seems to have been rooted in the late Ironworking traditions. The single coin from Masakasa is a vivid example, while clearly intensively used crucibles are plentiful in these sites. A great number of crucibles for smelting copper, as attested to by oxides adhering to the interior (Chittick 1966: 15), were recovered at Kilwa Island but fewer at Mafia during period II (12th – 13th c AD). Only a few of them were recovered at Rushungi in this work, which is not located far to the south of Kilwa. Crucible occurrences and uses were at a much higher frequency at Mnangole, where hundreds of pieces of intensively used crucibles with clear greenish cuprous staining on the interior were collected from the sequence. A date for a horizon bearing the crucibles and copper coins at Kilwa Island was dated to AD 1135 ±110 (Chittick 1966: 12). This date is of exactly the same order to those obtained by this work, representing the Proto-Swahili phase. From Chittick's (1974) discussions and illustrations up to this stage, it is becoming apparent that the Swahili ware tradition has its roots in this PSW culture of the southern coast, rather than the extant linguistic suggestions of northern coast origin (Spear 2000, Chami 1998, 2009b). Further research is necessary before coming to such a conclusion.

The Beginning of Settled Village Life.

The transition from the nomadic hunting and gathering mode of life to a settled food-producing tradition is still a complex matter among many scholars. Several models have been established to explain the problem. The migration model has remained the dominant model, influencing most archaeological interpretations made on the subject.

In sub-Saharan Africa, as widely accepted, people are said to have been exclusively foragers who relied on hunting, gathering and fishing for a very long time before AD/BC changeover. In the northern half of Africa cultivation and animal keeping was done by people who had not yet acquired metallurgical skills (Phillipson. 2005: 165). It is reasonably fair to assume that foragers in the southerly latitudes had developed metallurgical skills in order to better cope with the woodland forest of the region, but we still lack supporting evidence. Direct evidence for agriculture is still missing or very rare, even in most of the alleged agricultural EIW sites. Pollen analysis, which could better provide evidence of agriculture, faces many constraints in this region, especially based on the unconsolidated sandy soil nature of the coast and ever-changing prevailing winds. This work shows settled people who possessed pottery-making knowledge without metallurgical skills, identified by the PIW ware. They had settled on the shore, hence they would have fished. Such people existed on the Islands at the same time (Chami 2009b). Although this work could not come up with clear evidence concerning the latter, some of the ancient documents such as *Periplus of the Erythrean Sea* and *Ptolemy geographia* wrote about inhabitants of the coast having dugouts, sewn boats and fishing baskets (Freeman-Grenville 1966). Their settlements were located strategically not very near to the shore and shellfish exploitation is evident in most of their sites. Egyptian and greek records also provide simillar evidence for settled farming and animal domestication for this time period (Chami 2006, 2009b)

The distinction between gathering and hunting on one hand, and cultivation on the other is far from clear on the archaeology of this coast (Phillipson 2005:170). It now appears highly probable that the emergence of fully developed cultivation and animal herding economy was a result of a far longer period of intensive exploitation and experimentation, rather than a changeover from one mode to another. Domesticates also spread in that peoples migration. Exercising control over wild varieties of plant or animal already present in the area concerned seems to have played a much greater part in the transition towards most African cultivation. The hunting, gathering and farming seem to have continued to play an important part in most pre-colonial African economies. In few areas hunting and gathering had been providing a level of nutrient higher than achieved by the farming people (Lee 1968; Brooks et al. 1984)

To suggest pure migration or diffusion as the mechanism for the changeover from the LSA to PIW and later to the EIW settled life seems to be an oversimplification. It has been argued that the beginning of settled life communities represents economic adaptations over a very wide area that parallels a similar process of environmental change (Barich & Gatto 1997, Gremaschi & di-Lernia 1999), which was not an abrupt process but rather a gradual shift in ecological zones that intensified much later on Southern regions of Africa. At least in northern parts of Africa and the Nile valley, there had been for some millennia a tradition of intensive exploitation of various food resources, including aquatic resources. Sometimes these circumstances encouraged prolonged settlement on a particular site (Phillipson. 2005:159), a cultural phase which has not been verified on the southern regions, probably because of the absence of evidence like stone or brick constructions; or due to the prevailed wetter period that was more intense in the sub-Saharan Africa from ~11.7 to ~4.0 thousand years ago (Lonnie et al. 2002; Maley 1993).

In tropical rainy environments, rich in termites, there has until recently been a tradition of using wattles, brushes and mud for constructing living places. This tradition, which could have been rooted in the long prehistory of the region, is highly susceptible to weathering without leaving a trace. In a woodland environment, termites could

160

consume the polls and refill the postholes thoroughly with soil that blend with the surroundings within a few rainy seasons. Otherwise larger parts of the woodlands coastal and hinterlands were too wet for a permanent settlement. Probably these could reasonably be used to explain why we do not find structures related to living spaces.

Cultivation once adopted did not necessarily lead to a rapid abandonment of old traditions in many areas of Africa. At some point, farmers seem to have reverted to hunting and gathering, especially during agricultural failure or at interludes. Both activities continued to play an important part in most early, but also some of the recently African economies. On the other hand, in several African societies still domestic animals serve as embodiments of wealth and indicators of social status more than cultivation. Whether this explains the observed lack of evidence for intensive agriculture is yet to be decided, but it might have been rooted in the history of the region.

Given the above explanation on the subject matter, the conception of Bantu migrations with their attributed EIW traditions has been offered as an outstanding paradigm that explain the shift to a new mode of settled food production (Phillipson. 2005, 1976, Huffman 2005, 1970, Mitchell 2002, Posnansky 1961, Murdock 1959) The transition to food production is still widely thought of in archaeological circles as having universally entailed the immigration of speakers of new languages, who brought with them new subsistence techniques and material and cultural repertoires. However, This explanation, of the establishment of the EIW culture that was assumed to entail the oldest evidence of settled life, cultivation, metallurgical skills and pottery making in central, coastal and southern Tanzania, and further south of the region, was a gross oversimplification (see Soper 1971:8).

The mass of growing archaeological data is now entirely against the migration model. MSA-derived traditions are now evidently known to have existed even along the shores and islands of East Africa (Chami 2009b). The woodlands of South-eastern Africa, previously conceived as a tsetse fly barrier for a settled early farming tradition (Clark 1970), are apparently known to have been a foraging zone for the MSA and succeeding Stone Age communities. The tsetse problem could rather be viewed as a recent phenomenon. Communities with a settled way of life and pottery making could have been well established on the coast and its interior woodland since the beginning of the last millennium BC or even earlier, more than 3000 years ago. Their culture is herein identified as PIW, to include the one practised by settled "tillers of the soil" with little or without knowledge of iron smelting. Most scholars have rejected the time for the 'sewn boats' and 'fishing baskets' mentioned in the Periplus documents (Casson 1989). In other words, people of the coast had been fishing, exploiting woodland resources and perhaps cultivating long before the EIW times. The woodland is a sorghum-millet zone, part of the area suggested by linguistic study to have been settled before the spread of EIW communities (Oliver, 1966). In Southern Tanzania, this is also an area with evidence of pre-EIW settlements (Fig.7.5). Their fishing gears and houses must have been of plants and wood products. It is mostly the pottery

that has survived, and that is the PIW pottery discussed in this work with Mnaida
tradition described as one variant of the PIW culture. Shellfish from seashores, lakes,
or rivers are said to have formed an important part of the prehistoric diet for many
thousands of years (Waselkov 1987). Huge shell mounds and accumulations of
abandoned shells are common in many parts of the coast worldwide (Cook 1946,
Cook & Treganza 1950, Msemwa 1994). Around Lake Victoria, for instance, some
of the sites have proved associations of shell-mounds with LSA and others with
EIW farming traditions (Lane 2004: 257). There are evidence of the exploitation of
marine resources among the coastal communities of the early first millennium AD in
southern Mozambique and Natal in South Africa.

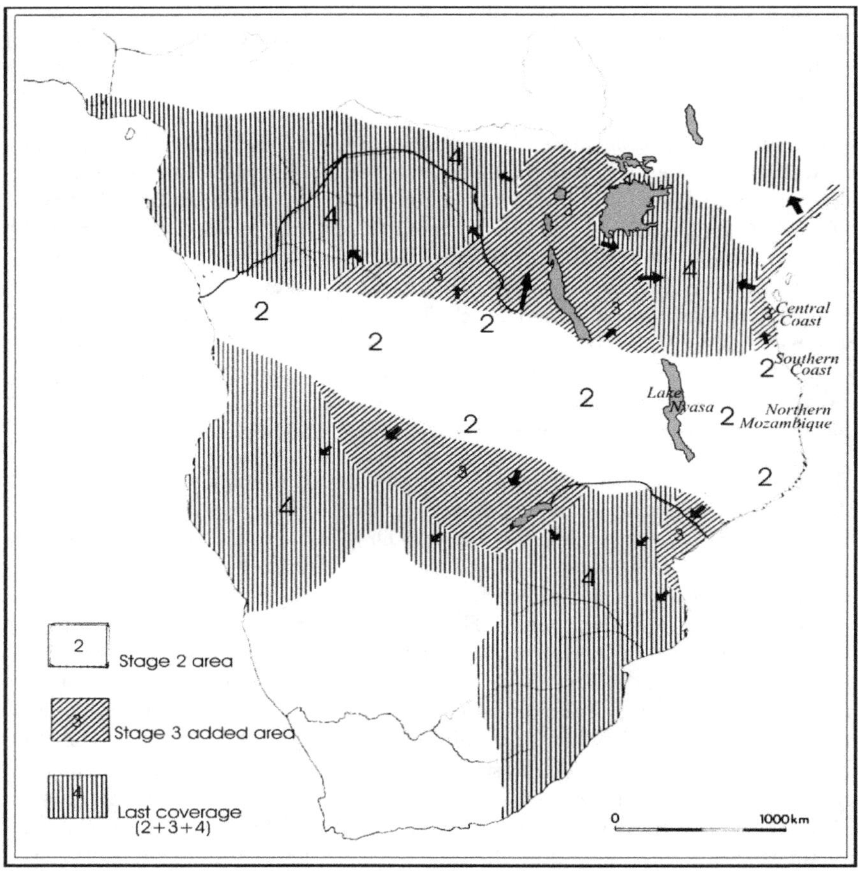

*Fig.7.5: The woodland region (stage 2) settled by early farming communities who adopted and
developed Indonesian crop cultivation according to linguistic data (Oliver 1966:369) modified
and substantiated by the PIW data from southern Tanzania. This is the region that also bears very
similar EIW pottery designs.*

Just like on the southern coast of Tanzania, agriculture is less well known due to the absence of primary evidence in the archaeological record, but sorghum and millet have been grown since time immemorial , and sites of early farming based on early ironworking ceramic styles are seen to occur in areas of high productivity (Sinclair 1993: 426). The latter may suggest extensive use of once fertile land before becoming exhausted. If that is correct, fishing and farming may have supplemented one another.

The Settlement Pattern.

It has been learnt that on most of the southern coast, the older sites are located far away from the shore, and higher up the hills away from rivers or water sources. Initially it was assumed that these sites would be found close to modern settlements, which are now on the shores of the Lake Kitere, Mikindani bay, Mchinga bay, Kiswere bay and so on. Among the recent phenomena along the shore is the formation of shell mounds resulting from the exploitation and processing of shellfish on the shore. Some of the researchers on shellfish exploitation were deceived by this modern behaviour and assumed that similar patterns would be found in the archaeological past (Shawcross 1972, Msemwa 1994,). Ironically, most of these earlier sites are located on now unfertile areas where the soil is exhausted and with stunted vegetation far from the valleys and water bodies.

During excavations, cultural gaps were noted in most of the sequences, although small. Mnaida and Nakatumbatu sites on one hand and Kabisela on the other, clearly demonstrate this situation. While the oldest sites at Mnaida dates to the last millennium BC, on its eastern and northwestern, side there are sites of different ages. To the northwest it begins with the TIW, while to the north it starts with typical Plain ware phase, after which the site was abandoned until recently. On the opposite side of Kabisela, sites begin with the PSW phase, while at Pemba, which is near these settlements, begun with the EIW. The population of the PIW was not so big but quite distinctively clustered as villages as demonstrated in the previous chapters.

Further towards the hinterland, Kitere settlements offered another example. The river basin with renewable fertility and the lake has only been recently settled. The oldest settlement sites are high up the hill with steep slopes, which rises to more than 300 metres within a short distance. They seem to have ignored the extensive fertile basin beside the river as well as around the lake, although their daily lives would have highly depended on it. It is becoming apparent that the valleys were too wet to be settled permanently.

The observed patterns could be explained in three different ways: the Rainfall could have been too much, hence the risk of floods prevented people from settling along the valleys; or the water table could have been too high to cause swampy conditions on the lowlands. In addition, predictable low security especially from attacks from the sea could have made people to hide their dwellings on the hills. An ethnographic study conducted among modern settlers showed different views. Some of them said they settled near the shore due to their frequent involvement in maritime activities. Others believed that their presence close to the shore is because of social connections with the historic town, assuming it to be the oldest settlement in the area. The rest believed it is the consequence of the National Settlement Adjustment Programme, which re-concentrated most of the population in new villages and banned 'hopping' behaviour.

Climatic conditions of the region suggested a prolonged wetter and warmer conditions that prevailed throughout the early and Middle Holocene, an episode that is clearly recorded by the Kilimanjaro summit Ice field (Lonnie et al. 2002). The local climatic conditions of the Southern coast suggested prolonged but fluctuating drier conditions during the late Holocene, with a single heavy rainy season that coincides with the months with the highest temperature (in Chapter Four). Few cultivation seasons were enough to upset the natural balance of the soil fertility, leading to a decline in the yield. The clear preference for particular settlement locations shown by early farmers 'hopping' from one favourable location to another during the late Holocene, with intervening areas settled only much later, would have been contributed by the loss of soil fertility. Hickman's (1965) discussion on the 'problems' of the soil and its use in the Makonde area of southern Tanzania, described the Makonde style of shifting or rotational agriculture. The Makonde would change crops from year to year before allowing the ground to lie fallow to develop thicket for several years, after which it was cleared again, i.e. thicket operation (ibid. 23).

The question as to why these early settlers preferred to settle in such remote areas away from water sources was also approached ethographically. The first assumption was that people could have settled more comfortably in the fertile valleys with higher water retention. Rice cultivation could have been conducted in these wetter valleys as it is today. It was later learned that, among the ancient Makonde there was a taboo that restricted areas where it was appropriate to settle. Most Makonde, who formed the majority of the population in southeastern Tanzania, believed that nature advised their ancestors not to live in the valleys or near the 'great waters' as these were abodes of sickness and death (Gillman 1945). With the current observations, one may be compelled to believe that settlement patterns during the PIW tradition were greatly influenced by this cultural taboo. In that line of argument, rice should have been introduced in that region more recently.

Early settlements, until the time of the Plain ware tradition were mostly located at a corner or on the edge of open, arable land suitable for sorghum-millet cultivations, which has now been abandoned due to degradation. This argument puts into account the introduction of maize and cassava later during the colonial times. The humus colloids that used to hold the crumbs together in the soil have been used up and perhaps some of them were lost due to prolonged period of drought circles. Throughout the excavations, the observed colour change down the sequences was insignificant. All the excavations had to be conducted based on the arbitral levels. The humus layer is almost missing or immature in most of the sites. This suggests intensive use of the soil through tilling of the surface, associated with the leaching process, as suggested, before the EIW period. In other words, the land had been cultivated with minimum disturbance of the cultural sequences below for a long time. This could have been achieved through slash and burning agriculture, which is still a common practice in the region.

The startling clustering of PIW sites in the southern woodlands of Tanzania, and especially on the Southern coast (Fig.7.6), suggests long established contacts and interactions between the coast and the interior, which seems to have developed a more-or-less similar agricultural revolution later on during the EIW period. The stronger development of the Bantu language along the axis of their interaction has been argued by linguistic studies (Guthrie, 1962; Wrigley, 1960; Oliver, 1966) but rejected in favour of the migration model. A review of the early pottery that appears below or within the lower Bambata (Mazel, 1992) in the far interior of southern Africa, which is here presumed to be a distal end of the coastal-interior corridor of active interactions; could provide a better link of the PIW with the inland traditions. This work has demonstrated similar links during the EIW period, extending from the coast into slightly south-western interior (Fig.7.6).

Conclusion

This work has tried to resolve some of the extant conceptual problems concerning the peopling and cultural processes of the southen coast of Tanzania. As argued above, it may have not fully solved all the problems as everyone would wish. Some of the problems seem to be deeply rooted in the colonial paradigmatic framework of negative thinking about indigenous settlers of Africa (Chami 2006: 50-52, 2009a). Migration model has been mostly used to explain cultural changes Africa. Some of the explanations about the early settlers and their culture have been tested by this work and the results suggest long Holocene cultural community coupled with different periods adaptations to different natural and social economic conditions. The culture process of the ealy Holocene period remains to be made clear in the region of may study, but for the later period of Holocene the imaged obtained is solid, only subject to future scibintific tests.

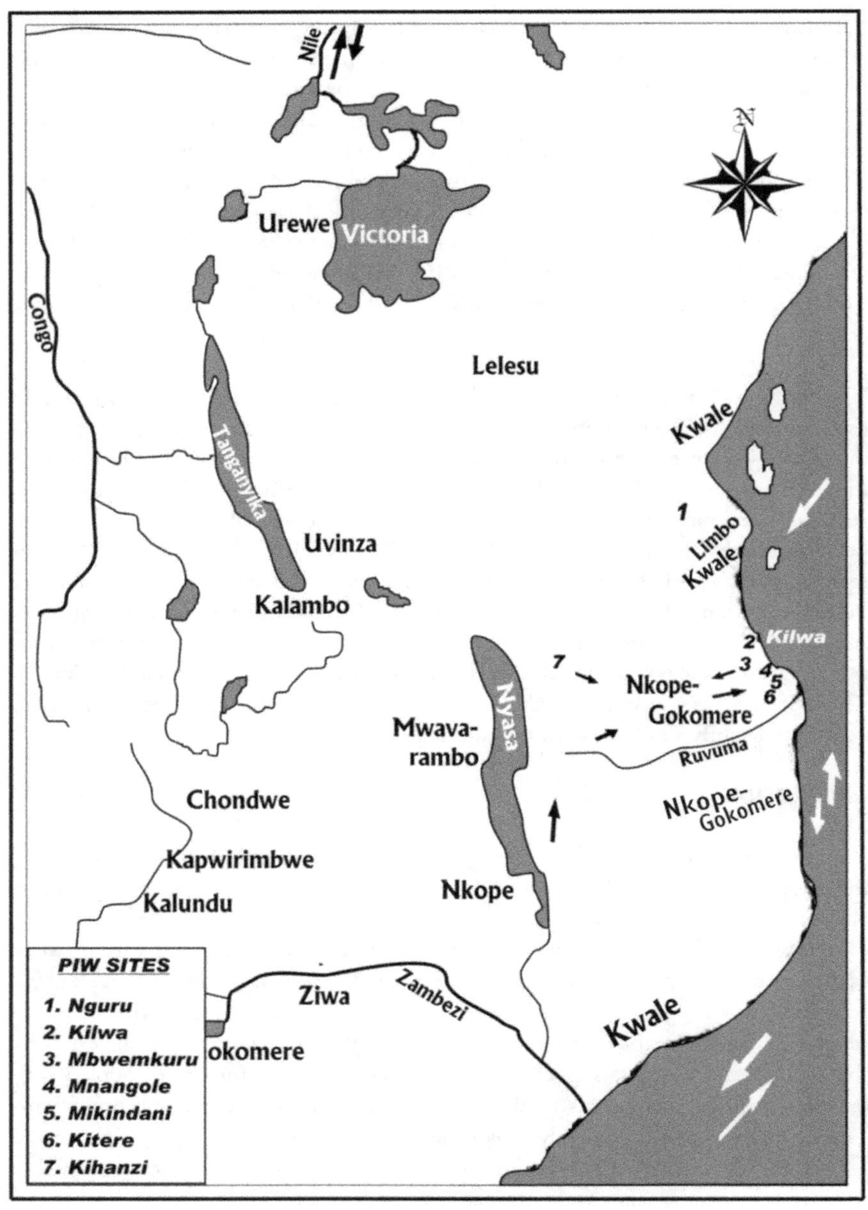

Fig. 7.6: Distribution of EIW variants in eastern and southern Africa and described PIW settlement sites.

166

REFERENCES

Abungu, G.H.O. (1989), Communities on the river Tana, Kenya: an archeological study of relations between the delta and river basin, AD 700-1890, Ph.D. thesis, University of Cambridge.

Abungu, G.H.O. (1994-95), Agriculture and settlement formation along the East African coast, Azania 29-30: 248-256.

Aitken, W.G. (1961), "Geology and Paleontology of the Jurassic and Cretaceous of southern Tanganyika", Geology Survey of Tanganyika 31 (1961)

Akazawa, T & C.M. Aikens (1986), "Prehistoric Hunter-Gatherers in Japan", Tokyo: University Museum Bulletin 27.

Allen, J. de V. (1993), Swahili Origins, London: James Curry.

Ambrose, S.H. (1984), "The introduction of pastoral adaptations to the highlands of East Africa", in J.D. Clark and S.A. Brandt (eds.), From Hunter to Farmer: The causes and consequences of food production in Africa. Berkley: University of California Press, pp. 212-39.

Ashmore, W. & R.J. Sharer (2006), Discovering our Past: a brief introduction to archaeology, 4th ed., New York: McGraw-Hill, Inc.

Barich, B.E. & M.C. Gatto (1997), Dynamics of Populations: Movements and Responses to climatic Changes in Africa. Rome: Bonsignori.

Barthelme, J. 1985. Fisher hunters and Neolithic pastoralists in East Turkana, Kenya. Cambridge: BAR International sewes 254.

Baxter, H.W (1944), "Pangani: The Trade Center of Ancient History", Tanganyika Notes and Records 17, pp. 55-64.

Berger, A. & M.P. Loutre (2002), An Exceptionally long interglacial Ahead? Science 297 (5585): 1287-1288.

Berry, L. (1971), "Hydrology of Major rivers", in Tanzania in maps, Tanzania Publishing House, pp. 24-25.

Binford, L.R. (1962), "Archaeology as Anthropology", American Antiquity 28, pp.217-25.

Binford, L.R. (1964), "A Consideration of the Archaeological research design", American Antiquity 29, pp.425-41.

Binford, L.R. (1969), "Some comments on Historical versus Processual Archaeology", Southern Journal of Anthropology, 24, pp. 267-75.

Binford, L.R. (1972), An archaeological perspective. New York: Seminal Press.

Binford, L.R. (1983a), In Pursuit of the Past: Decoding the Archaeological Record. London: Thames and Hudson.

Bolliger, B. 2005. Acient Civilization in a Eqypt. Willoughby: Global Book Publishing.

Binford, L.R. (1983b), Working at Archaeology. Orlando, FL: Academic Press.

Bower, J.R. & C.M. Nelson (1978), "Early pottery and pastoral cultures of the rift valley", Man 13, pp.554-66.

Bower, J.R. & Chadderdon, T.J. (1986), "Further excavations of Pastoral Neolithic sites in Serengeti", Azania 21, pp.129-133.

Bower, J.R.F. et al. (1977), "The University of Massachusetts Later Stone Age/Pastoral Neolithic comparative study in central Kenya", Azania 12, pp.119-46

Brooks, A.S; D.E. Gelburd & J.E.yellen (1984), Food production and culture change among the !Kung San: implications for prehistoric research, in J.D. Clark & S.A. Brandt (eds.), From Hunters to Farmers: The causes and Consequences of Food Production in Africa, pp. 293-310. Berkeley: University of California Press.

Burkill, I.H. (1953), "Habits of man and the Origin of Cultivated Plants of the Old World", Proc. Linnean Society of London 164, pp.12-41.

Campbell, B.G & J.D. Loy (2000), Humankind Emerging, 8th edit. Ally and Bacon: Pearson Education Company

Casson, L. (1989), Periplus Maris Erythraei, Princeton: Priceton University Press.

Chami, F.A. (1992), "Limbo. Early iron –working in south-eastern Tanzania", Azania 27, pp. 45-52.

Chami, F.A. (1994), The Tanzanian Coast in the First Millennium AD, Uppsaliensis SAA 7, Uppsala.

Chami, F.A. (1996), "The excavation of Kiwangwa Late Stone Age site", In Pwiti, G. & R. Soper (eds.), Aspects of African Archeology. Papers of the 10th Congress of Pan Afr.Assoc.Preh&Rel. Studies; University of Zimbabwe, Harare, pp.307-316.

Chami, F.A. (1998), "A Review of Swahili Archeology", in African Archeological Review 15(3), pp.199-218.

Chami, F.A. (1999a), "Greco-Roman Trade link and the Bantu migration theory", Anthropos 94 (1999), pp.205-215

Chami, F.A. (1999c), "The Early Iron Age on Mafia Island and its relationship with the Mainland", Azania 34, pp.1-10.

Chami, F.A. (2002a), "Kaole and the Swahili World", in Chami, F & G.Pwiti (eds.), Southern Africa and the Swahili World, Studies in the African Past, 2, pp.1-14.

Chami, F.A. (2002b), "People and Contacts in the ancient western Indian Ocean Seaboard or Azania", Man and Environment 27(1), pp.33-44.

Chami, F.A. (2004), "The archaeology of Mafia Archipelago, Tanzania: New evidence for Neolithic Trade links", in F. Chami, G. Pwiti and C. Radimilaly (eds.), Studies in the African Past, 4, pp.73-101.

Chami, F.A. (2005), The Archeology of Pre-Islamic Kilwa, (unpublished) report submitted to the Department of Antiquity for the Kilwa Tanzania/French sponsored Project, March 2005.

Chami, F.A. (2006), The unity of African Ancient History 3000BC to AD 500. Mauritius, E & D. Publishers.

Chami, F. (2009a) - Atomic model in the Study of Africa Past. In Schmidt, P. ed, *Post colonial archaelogies*, Page, Santale.

Chami, F ed. (2009b) Zanzibar and the Swahili Coast from 30,000 years ago. Dar es Salaam: E&D Vision Publishing Ltd.

Chami, F. (2011) - The problem of equifinality in archaelogy. *Studies in the Africa Past* 9: 1-12

Chami, F. & A. Kwekason (2003), "Neolithic Pottery Traditions From the Islands, Coast and the Interior of East Africa", African Archeological Review 20(2), pp.65-80

Chami, F. & R. Chami (2001), "Narosura Pottery from the Southern Coast of Tanzania: First incontrovertible coastal Late Stone Age pottery", Nyame Akuma 56, pp.29-35

Childe, V.G. (1925), The Danube in Prehistory, London: Routledge & Kegan Paul.

Childe, V.G. (1951), Social evolution, London: Watts and Co.

Chittick, N. (1965), "The 'Shirazi' Colonization of East Africa", Journal of African History 6, pp.275-95.

Chittick, N. (1966), "Kilwa: a preliminary Report", Azania 1, pp.1-37

Chittick, N. (1974), Kilwa: An Islamic Trading city on the East African coast (2 vols.), the Finds, British Institute in Eastern Africa, Nairobi.

Chittick, N. (1975), The peopling of the East African Coast, in N. Chittick & R. Rotberg (eds.), East Africa and orient, London: Holmes & Meier Publishers, pp.16-43.

Chittick, N. (1984), Manda: Excavation at an Islamic Port on the Kenya Coast, Nairobi: Brit. Inst. E. Africa.

Clark, J.D. (1962a), "Africa South of Sahara", in Braidwood & Willey (eds.), Courses Towards Urban Life. Viking Fund Publications in Anthropology 2.

Clark, J.D. (1962b), "The Spread of Food-production in Sub-Saharan Africa", Journal of African History 3, pp.211-28

Clark, J.D. (1967), "The problem of Neolithic Culture in Sub-Saharan Africa", W.W. Bishop & J.D. Clark (eds.), Background to Evolution in Africa, Chicago: University of Chicago Press, pp.601-627.

Clark, J.D. (1970), "The Prehistoric Origins of African Cultures", in J.D. Fage & R.A.Oliver (eds.), Papers in African Prehistory, Cambridge: Cambridge University Press, pp.1-23

Clark, J.D. (1980), "Early human occupation of African savanna environments", in Harris, D.R. (eds.), Human Ecology in Savanna Environments, Academic Press, London, pp.41-71

Clarke, D.L. (1968), Analytical Archaeology, London: Methuen.

Clarke, D.L. (1972), "Models and Paradigms in Contemporary Archaeology", in Clarke, D.L. (ed.), Models in Archaeology, London: Methuen, pp.1-60

Clarke, D.L. (1978), Analytical Archaeology. 2nd revised ed. 1978. London: Methecen.

Collett, D.P. (1988), The spread of early iron producing communities in eastern and southern Africa, PhD thesis, University of Cambridge.

Collett, D. & P. Robertshaw (1983), "Pottery traditions of the early pastoral communities in Kenya", Azania 18 pp.107-125.

Cook, S. F. (1946), "A reconstruction of shell mounds with respect to population and nutrition", American Antiquity, 12 (1), pp.50-53.

Cook, S.F.& A.E. Treganza (1950), "The quantitative investigation of Indian mounds with special reference to the physical components to the probable material culture", University of California publications in American Archaeology and ethnology, 40 (5), pp.227-62.

Coon, C. (1965), The Living Races of Man, London: Jonathan Cape.

Cremaschi, M. & S. di Lernia (1999), "Holocene Climatic Changes and Cultural Dynamics in the Libyan Sahara", African Archeological Review 19, pp.211-238.

Crumley, C. (1994), Historical Ecology: Cultural Knowledge and Changing landscape, Sante Fe (Nn): SAR Press.

Cruz-e Silva. (1977), "First indication of early Iron Age in the southern Mozambique: Matola IV 1/68", in R.E. Leakey & B. Ogot (eds.), Proceeding of the 8th Pan African Congress of Prehistory and Quaternary Studies, Nairobi: Intern. L.L. Mem. Inst. of Africa Prehistory, pp.349.

Dalby, David (1975), "The prehistoric implication of Guthrie's comparative Bantu", Journal of African History 16 (1975), pp.481-501.

Datoo, B. (1975), Port Development in east Africa, Dar es salaam: East African literature Bureau.

Davidson, B. (1964), Old Africa Rediscovered. London: Gollancz

Dunnel, R.C. & W.S. Dancey (1983), "The sites-less Survey. A regional scale Data collection", Advances in Archaeological Method and Theory 6, pp.267-288.

Ehret, C. (1982), "Linguistic Inference About Early Bantu History", in C. Ehret & M. Posnansky (Eds.), The Archeological and Linguistic Reconstruction of African History. Berkeley: University of California Press.

Ehret, C. (1998), An African Classical Age, Charlottesville, University Press of Virginia.

Ehret, C. (2002), The Civilization of Africa: a history to 1800, Charlottesville, University Press of Virginia.

Ellen, J. de V. (1982), "Settlement Pattern on the East Africa Coast, ca. AD 800-1900", in Leakey, R.E. & Ogot, B.A (eds.), Proceedings of the 8th Pan African Congress of Prehistory and Quaternary Studies, Nairobi. The Inter. L. Leakey Memorial Inst. for African Prehistory, pp.361-363

Fagan, B.M. (1961), "Pre-European iron working in Central Africa with special reference to Zambia", Journal of African History.2 (2), pp.199-210.

Fagan, B.M. (2005), In the Beginning: An introduction to Archaeology. New Jersey: Pearson-Prentice Hall.

Flannery, V.K. (1968), "Archaeological Systems Theory and Early Mesoamerica", in B.J Meggers (ed.), Anthropological Archaeology in the America, 142 pp.67-87.

Freeman-Grenville, G. (1962), The Medieval History of the Coast of Tanganyika. London: Oxford University Press.

Freeman-Grenville, G. (1966), The East African Coast. London: Oxford University Press.

Freeman-Grenville, G. (1975), The East African Coast: selected documents from the first to the earliest nineteenth century. Oxford: Clarendon Press.

Gibbon, G. (1989), Explanation Archaeology. London: Oxford University Press.

Giddens, A. (1979), Central Problems in Social Theory, London: Macmillan

Giddens, A. (1982), Profile and Critiques in Social Theory, London: Macmillan.

Gillman, H. (1945), "Bush fallowing on the Makonde Plateau", Tanganyika Notes and Record 19, pp.34-44.

Guthrie, M. (1962), "Some developments in the prehistory of the Bantu languages", Journal of African History 3(2) pp.273-82.

Harding, J. (1961), "Late Stone Age sites on the coast of Tanganyika coast", MAN, vol. LXI, No. 221, 1961.

Hathout, A.S. (1983), "Agricultural soil capacity of southeast Tanzania", in Soil Atlas of Tanzania. Dar es Salaam: Tanzania Publishing House

Hawkes, J. (1968), "The proper Studies of Mankind", Antiquity 42, pp.255-62

Hester, T.R; H.J. Shaffer, & K.L Feder. (1997), Field Methods in Archaeology.7th ed., New York: McGraw-Hill.

Hickman, G.M. (1965), The lands and people of East Africa, 4th ed., London: Longman, Green & Co. Ltd

Hodder, I. (1986), Reading the Past: Current Approaches to Interpretation in Archaeology. Cambridge University Press.

Hodder, I. (1999), The Archaeological process: An introduction. Oxford: Blackwell.

Hodder, I. & C. Ortorn, (1976), Spatial analysis in Archaeology. Cambridge: Cambridge University Press.

Holl, A. (1993), "Transition from LSA to Iron Age in the Sudano-Sahelian zone: a case study from the perichadian plain", in T. Shaw et al., (eds.) The Archeology of Africa: Food, metal and town, London: Routledge, pp.330-43.

Horton, M. (1984), The Early Settlements of the Northern Swahili Coast, Ph.D. thesis, Cambridge University, Cambridge.

Horton, M. (1990), "The Periplus and East Africa", Azania 25, pp.95-99.

Horton, M. (1996a), Shanga: The archaeology of a Muslim trading community on the coast of East Africa London: British Institute of Eastern Africa.

Horton, M. (1996b), "Early Maritime trade and settlement along the coast of Eastern Africa", in J. Reads (ed.) The Indian Ocean in Antiquity London: Kegan Paul, pp.439-60.

Huffman, T.N. (1970), "The Early Iron Age and the Spread of Bantu", South African Archeological Bulletin 25, pp.3-21.

Huffman, T.N. (1976), "Gokomere pottery from the Tunnel site, Gokomere", South African Archeological Bulletin 31, pp.31-53

Huffman, T.N. (1980), "Ceramics, Classification and Iron Age entities", African Studies 39 (2), pp.123-74.

Huffman, T.N. (1989), Iron Age Migrations: The Ceramic Sequence in the Southern Zambia, Excavation at Gundu and Ndonde. Johannesburg: Witwatersrand, University Press.

Huffman, T.N. (2005), "The stylistic origin of Bambata and the spread of mixed farming in southern Africa", Southern African Humanities 17, pp.57-79.

Huffman, T.N. & R. Herbert (1994-95), "New Perspectives on eastern Bantu", Azania 29-30, pp.27-36

Huntingford, G. (1963), "The Peopling of the interior of the East Africa by its modern inhabitants", in R. Oliver & G. Mathew (eds.), History of East Africa, Oxford: Clarendon Press, pp.58-93.

Huntingford, W.B. (1980), The Periplus of the Erythreaean Sea, London: The Hahluyt Society.

Isaac, G. (1974), "Stone Age Remains on Kilwa Island", in N. Chittick, Kilwa: an Islamic Trading city on the East African coast, Vol. II, Nairobi: British Institute in Eastern Africa, pp.254- 56.

Johnson, M. (1999), Archaeological theory: An introduction. Blackwell Publishers, LTD, UK.

Jones, H. (1960), (Trans.) The Geography of Strabo, Vol. I. London: William Heinemann Ltd.

Juma, A 2004 Unguja Ukuu on Zanzibar. Uppsala: Studies in Global Archaelogy

Juma, A. & Lofgren, (1992), "A site guide to Unguja Ukuu and the fieldwork done in 1990", in Sinclair, P. & A. Juma (eds.) Urban origins in Eastern Africa, proceedings of the 1991 workshop in Zanzibar, Stockholm. The Swedish Central Board of National Antiquities, pp.10-14.

Karoma, J. (1982), "New evidence of Paleohistoric human Activities along the Kilwa coast, southern Tanzania", University of Dar es Salaam (Unpublished report).

Kessy E, 2009. Analysis of lithic artetact, from Kuumbi and Mwanampambe Caves, In Chami, f. ed: Zanzibar and the Swahili coast from c. 30,000 years ago. Dar es Salaam : E&D Vision Publishing Ltd.

Kirkman, J.S. (1958), "Kilwa: The cutting behind the defensive wall", Tanganyika Notes & Records 50, pp.94

Knutsson, K. 2007. Preliminary analysis of Lithics from Kuumbi Cave, Zanzibar. Studies in the African Post 6:8-17

Kusimba, C.M. & S.B. Kusimba (2005), Mosaics and interactions: East Africa, 2000BP to the present. In Stahl, A.B. (ed.), African Archaeology: A critical introduction, pp 392-419. Oxford: Blackwell.

Kwekason, A.P. (2002), "Geo-environmental aspects of the Dar es salaam Area", in F. Chami & G.Pwiti (eds.), Studies in the African Past, 2, Dar es Salaam: Dar es salaam University Press, pp.15-24

Kwekason, A.P. (2007), "Pre-early Iron Working Sedentary Communities on the Southern Coast of Tanzania", in G.Pwiti, C.Radimilary & F.Chami (eds.), Studies in the African Past 6, pp.20-40.

Kwekason, A. and F. Chami 2003. Archaeology of Muleba, Southwest of Lake Nyanza. *Studies in the African* Post 3: 59 - 85.

Lacroix, W.F. (1998), Africa in Antiquity: a linguistic and toponymic analysis of Ptolemy's map of Africa. Saarbrucken:Verl. Fur Entwicklungspolitic.

Lane, L.P. (1945), "To Utete by ship", Tanganyika Notes and Records 20, pp.55-59.

Lane, Paul (2004), "The 'moving frontier' and the transition to food production in Kenya", Azania.39, pp.243-64.

Larsen, C.S. (2008), Our Origins: discovering physical anthropology. New York: W.W. Norton & Co. Inc.

Lee, R.B. (1968), What hunters do for a living, in R.B. Lee & I. DeVore (eds.), Man the Hunter, pp.30-48. Chicago:Aldine.

Lonnie, G.T; E.M. Thompos; M.E. Davis; K. Henderson; H. Brecher; V. Zagorodnov; T. Mashiotta; P. Lin; V. Mikhalenko; D. Hardy; J. Beer (2002), Kilimanjaro Ice Core Record: evidence of Holocene climate change in tropical Africa; Science 298: 589-593.

Maley, J. (1993), "Climatic and vegetation history of the equatorial regions of Africa during the upper Quaternary", in Shaw, T. et al (eds.), Archaeology of Africa: Food, Metal and Towns, London: Routledge, pp.43-52.

Mathew G. (1963), "The East African Coast until the coming of the Portuguese", in Oliver, R & G. Mathew (eds.) History of East Africa, Oxford: Clarendon Press, pp.94-168.

Mathew G. (1975), "The Dating and the Significance of the Periplus of the Erythrean sea", in Chittick, N. & R. Rotberg (eds.), East Africa and the Orient, New York: African Publishing Co., pp.147-63.

Mazel, A. (1992), "Early Pottery from the eastern parts of southern Africa", Southern African archaeological Bulletin 47, pp.3-7.

McClanahan, T. (1996), "Oceanic ecosystems and pelagic fisheries", in T. McClanahan and T. Young (eds.), East African Ecosystems and their conservations, Oxford: Oxford University Press pp.39-66.

Mehlman, M.J. (1977), "Excavations at Nasera Rock", Azania 12, pp.111-118.

Mehlman, M. (1979), "Mumba-Höhle revisited: The relevance of a forgotten excavation to some current issues in East African Prehistory", World Archeology 2(1), pp.81-94.

Mehlman, M.J. (1989), Late Quaternary archaeological sequences in northern Tanzania, PhD thesis, University of Illinois, Urbana.

Miller, D. & N. J. van der Merwe (1994), Early metal working in sub-Saharan Africa: a review of recent research, Journal of African History 33,pp. 1-36.

Mitchell, P. (2002), The Archaeology of Southern Africa. Cambridge: Cambridge University Press.

Msemwa, P. (1994), An Ethno archaeological Study of Shellfish Collection in a complex urban setting. Ph.D. dissertation, Brown University, Providence. R.I.

Msemwa, P. (2001), "Archeology of upper Rufiji catchments", in Chami, F., G. Pwiti, C. Radimilahy (eds.), Studies in the African Past 1, Dar es Salaam: Dar es Salaam University Press, pp.40-52.

Mueller, J.A (ed.) (1974), Sampling in archaeology. Tucson: University of Arizona Press.

Murdock, G.P. (1959). Africa: its People and their culture History. New York. McGraw Hill.

Ngusaru, A. (2002), "Geological history", in M.D. Richmond (ed.), A field guide to the seashores of eastern Africa and the western Indian Ocean Islands, Sida/ SAREC-UDSM, pp.12-21.

Nurse, D. and T. spear 1985. The Swahili. Philadefilia: Univesity Press.

Nyoni, E.N. (1998), Development Approaches and Woman economic Empowerment, M.A dissertation, University of Dar es Salaam.

Odner, K. (1972), "Excavations at Narosura, a Stone Bowl site in the southern Kenya highlands", Azania 7, pp.25-92.

Oliver, Roland (1966), "The problem of the Bantu Expansion", Journal of African History 7, (3), pp.361-376.

Omi, G, 1982. An Interim report on the East and Northeast Africa Pre-history research project 1982. Matsumoto Shinshu Univesity.

Orton, C. (2000), Sampling in Archaeology. Cambridge: Cambridge University Press.

Orton, C., P. Tyers, & A. Vince (1993), Pottery in archaeology. Cambridge: Cambridge University Press.

Phillipson, D.W. (1975), "The Chronology of the Iron Age in Bantu Africa", Journal of Africa History.16, pp.321-42.

Phillipson, D.W. (1976a), The Prehistory of Eastern Zambia. Nairobi: British Institute in Eastern Africa.

Phillipson, D.W. (1976b), "Archaeology and Bantu linguistic", World Archaeology, 8, pp.65-82.

Phillipson, D.W. (1977), The Later Prehistory of Eastern and Southern Africa. London: Heinemann.

Phillipson, D.W. (1985/1993), African archaeology. Cambridge: Cambridge University Press.

Phillipson, D.W. (2005), African Archeology, (3rd edit.), Cambridge: Cambridge University Press.

Posnansky, M. (1961), "Bantu genesis", Uganda Journal, 25(1), pp.86-93

Posnansky, M. (1962), "The Neolithic Cultures of East Africa", in Actes du 4e Congres Pan-Africain de Prehistoire et de L'Etude du Quaternaire, III, Tervuren, pp.237-81.

Radimilaly, C.(1992), "Travaux de reconnaissance l'arche'ologique a'mahilaka en 1989-1990", in Sinclair, P.J. & A. Juma (eds.), Urban Origins in Eastern Africa: Proceedings of the 1991 workshop in Zanzibar, Stockholm: The Swedish Central Board of National antiquities, pp.22-29.

Rawling, G. (1964), (Trans.) Histories of Herodotus, London: J.N. Dent & Sons.

Renfrew, A.C. (1989), "Comments to Archaeology into the 1990s", in Norway Archaeological Review 22(1), pp.33-41.

Renfrew, A.C. & P. Bahn (1996), Archaeology: Theories, Methods and Practice. 2nd ed. New York: Thames and Hudson.

Rice, P.M. (1987), Pottery Analysis: A sourcebook. Chicago: University of Chicago Press.

Robertshaw, P. (1990), Early Pastoralists of Southwestern Kenya. Nairobi: British Institute in Eastern Africa.

Sampson, C.G. (1979), "Comments on D. Stiles' Paleolithic culture and culture change: experiment in theory and method", Current Anthropology 20 pp.15-16.

Sassi, E, (2006), Re-excavation at Nguruni site in Kilwa Kisiwani, M.A (archeology) dissertation, University of Dar es Salaam.

Seddon, D. 1968. The Origins and Development of Agriculture in East and Southern Africa. Current Anthropology **9**(5): 489-506.

Schmidt, P.R. (1997). Iron Technology in East Africa: Symbolism, Science and Archaeology, Bloomington: Indiana University Press.

Schmidt, P.R. (1978). Historical Archaeology: a structural approach in an African culture. Westport: Greenwood Press.

Shanks, M. & C. Tilley (1987a), Re-Constructing Archaeology: Theory and Practice [RCA], Cambridge: Cambridge Univ. Press.

Shanks, M. & C. Tilley (1987b), Social Theories and Archaeology [STA], Cambridge: Polity Press.

Shanks, M. & C. Tilley (1992). Re-Constructing Archaeology: Theory and Practice (2nd edn.), London: Routledge

Shanks, M. & I. Hodder (1995), "Processual, Postprocessual and Interpretive archaeologies", in Hodder et al. (eds.) Interpreting Archaeology: Finding meaning in the past, London: Routledge. pp.3-29.

Shawcross, W. (1972), "Energy and ecology: thermodynamic models in archaeology", in David L. Clarke (ed.), Models in Archaeology. London: Methuen.

Shepard, A.O. (1971). Ceramics for the Archaeologist, 2nd ed. Washington, DC: Smithsonian Institute.

Simmonds, N.W. (1959), Banana. London

Sinclair, P.J. (1982), "Chibuene: An early trading site in the southern Mozambique", Paideuma 28, pp.149-64.

Sinclair, P.J. (1987), Space, Time, Social Formation. A territorial approach to the Archaeology and Anthropology of Zimbabwe and Mozambique, c.0-1700AD, Aun.9, Uppsala: Societies Archaeological Upsaliensis.

Sinclair, P.J. (2007), "What is the Archeological Evidence for External trading Contacts on the East African Coast in the First Millennium BC?" in J. Starkey, P. Starkey & T. Wilkinson (eds.), BAR International Series 1661, pp.187-194.

Sinclair, P.J. et al., (1992), "The impact of information technology on the Archaeology of Southern and Eastern Africa, the first decades", in Reilly, P & S. Rahtz (eds.), Archaeology and the information Age: A Global Perspective, London: Routledge, pp.30-40.

Sinclair, P.J. et al., (1993), "A perspective on archeological research in Mozambique", in T. Shaw, et al. (eds.), The Archeology of Africa: Food, metal and Town, London: Routledge, pp.410-31.

Smith, B.D.(1999), The emergence of Agriculture. New York: Scientific American library.

Smolla, G. (1962), "Steingerate vom Tendagurrn", in Actes do IVe congrès panafricain de préhistoire et de L'Etude du Quaternaire, sect. III, pre- et proto- histoire, Tervuren, (1962), pp.243-248.

Spear, T. 2000, Early Swahili history reconsidered. "The International Journal of Africa Historical Studies 33 (2): 257 - 290.

Soper, R.C. (1971), "A general review of Early Iron Age of Southern half of Africa", Azania 5, pp. 39-52

Soper, R.C. (1982), "Bantu Expansion into Eastern Africa: Archeological Evidence" in Ehret, C. & M. Posnansky (eds.), The Archeological and Linguistic Reconstruction of African History, Berkeley: University of California Press, pp.223-238.

Stein, P.L. & B.M. Rower (2006) Physical Anthropology, 9th edit. The McGraw-Hill company

Sutton, J.E. (1974), "The aquatic civilization of middle Africa", Journal of African History 15, pp.527-546.

Sutton, J.E. (1990), A Thousand Years of East Africa. British Institute in Eastern Africa, Nairobi.

Sutton, J.E. (1994-95), "East Africa: Interior and coast", Azania 29-30, pp.227-231.

Taylor, W.W. (1948), A Study of Archaeology. Menasha: American Anthop. Society

Thorp, C. (1992), "Archaeology in the Nguru Hills", Azania 27, pp.21-44.

Tilley, C. (1991), Material Culture and Text: The Art of Ambiguity. London: Routledge

Trigger, B.G. (1989a), A History of Archaeological Thought, Cambridge: Cambridge University Press.

Trigger, B.G. (1989b), "Comments on Archaeology into the 1990s" in Norwegian Archaeological Review.22, pp.28-31.

Trimingham, J.S. (1975), "The Arab Geographers and the East African coast" in Chittick, H.N. & R.J. Rotberg (eds.), East Africa and the Orient: cultural syntheses in pre-colonial times, New York: Holmes & Meier Publishers, Inc., pp.115-46.

UN-Habitat (2008), African cities at risk due to sea-level rise, United Nations Human settlements Program.

Waselkov, G. (1987), "Shellfish Gathering and Shell Midden Archaeology", in Advances in Archaeological Methods and Theory 10, pp.112-167.

Watson, P.J; S.A. Le Blanc & C.L Redman (1984), Archaeological explanation: Scientific Method in Archaeology 2nd ed. New York: Colombia University Press.

Wheatley, P. (1975), "Anacleta Sina-Africana Recenta", in Chittick, H.N. & R.J. Rotberg (eds.), East Africa and the Orient: cultural syntheses in pre-colonial times, New York: Holmes & Meier Publishers, Inc., pp.76-114.

Whiteley, W.H. (1951), "Southern Province Rock Paintings", Tanganyika Note and Records, pp.58-61.

White, F. (1983), The Vegetation of Africa. Parris: UNESCO.

Willcox, A.R. (1984), The Rock arts of Africa. Croom Helm Ltd., Provident House (GB).

Wright, H. 1984. Eary seafarers of the Comoro islands: the Dembeni phase of the IXth - Xth centuries. Azania XIX: 13-60.

Wrigley, C.C. (1960), "Speculations in the economic prehistory of Africa", Journal of African History 1 (1961), pp.201

Wylie, A. (1985), "The reaction against Analogy", in Advances in Archaeological Methods and Theories.8, pp.63-11.

Wylie, A. (1988). The New Archaeology: Tensions in Theory and Practice. Orlando, F2: Academic Press.

Zwernemann, J.J. (1983), Culture History and African Anthropology. Uppsala: Development of Cultural Anthropology.

<h1 style="text-align:center">APPENDEX I</h1>

<h2 style="text-align:center">Excavation details of the Southern coast (TZA) during 2006-2007.</h2>

Table 1.1 Mikindani cluster 2006

YEAR	Trench	Size	Depth-Range	Pottery Total	Décorrated	Shells	Bead	Bones	Lithic
2006	M1A	2x1	0 - 30	14	2	3	0	0	0
	M1B	2x1	0 - 30	10	0	7	0	0	0
	M1C	2x1	0 - 30	14	0	7	0	0	0
S/Total	**3**	**6m²**	**30cm**	**28**	**2**	**17**			
2006	M2A	4x1.5	0 - 40	156	14	80	0	0	0
	M2B	4x1.5	0 - 40	28	11	10	0	0	0
S/Total	**2**	**12m²**	**40 cm**	**186**	**25**	**90**			
2006	M3	3x3	0 - 50	305	72	0	2	0	0
S/Total	**1**	**9m²**	**50 cm**						
2006	M4	3x3	0 - 80	245	30	20	0	0	0
	M5	2x1	0 - 80	79	23	0	0		0
S/Total	**2**	**11m²**	**80cm**	**324**	**53**				
2006	M6	2x3	0 - 70	1261	234	18	10	48	0
S/Total	**1**	**6m²**	**70cm**						
2006	M7	1x1.6	0 - 50	658	2	0	0	0	0
S/Total	**1**	**1.6m²**	**50cm**						
TOTAL	**10**	**45.6m²**	**80cm**	**2770**	**388**	**145**	**12**	**48**	**0**

1.2 Rushungi cluster									
YEAR	**Trench**	**Size**	**Depth-Range**	**Pottery Total**	**Décorated**	**Shells**	**Bead**	**Bones**	**Lithic**
2007	T1	3x2	0 - 60	140	13	1190	2	0	4
	T2	3x2	0 - 40	178	8	284	0	0	9
	T3	4x2.5	0 - 30	37	4	0	0	0	0
	T4	3x4	0 - 40	862	134	0	49	0	0
TOTAL	**4**	**34m²**	**60cm**	**1234**	**159**	**1474**	**51**	**0**	**13**

1.3 Kitere (Mtwara)									
YEAR	**Trench**	**Size**	**Depth-Range**	**Pottery Total**	**Décor rated**	**Shells**	**Bead**	**Bones**	**Lithic**
2007	T1	2x1	0 - 60	55	5	0	0	0	22
	T2	2x2	0 - 60	102	24	0	0	0	72
TOTAL	**2**	**6m²**	**60cm**	**157**	**29**	**0**	**0**	**0**	**94**

1.4 Mnan'gole cluster									
YEAR	**Trench**	**Size**	**Depth-Range**	**Pottery Total**	**Décor-rated**	**Shells**	**Bead**	**Bones**	**Lithic**
2007	T1	3x1	0 - 50	1029	172	197	0	0	0
	T2	3x1	0 - 50	512	36	0	0	0	0
	T3	4x1	0 - 60	245	30	26	1	0	1
	T3E	2x1	0 - 60	186	14	2	0	0	3
	T4	3x1	0 - 90	475	48	0	1	0	6
	T5	2x2	0 - 80	2143	216	0	0	15	12
	T5E	2x1	0 - 80	598	48	0	0	5	13
	T6	2x1	0 - 70	203	20	0	1	0	0
	T7	2x1	0 - 40	148	27	5	0	0	0
	T8	2.5x1	0 - 60	172	26	0	0	0	1
	T9	3x1	0 - 80	294	12	0	0	0	4
	T10	2x1	0 - 50	72		153	0	5	0
TOTAL	**12**	**35.5m²**	**90cm**	**6077**	**649**	**383**	**3**	**25**	**40**

1.5 Mnaida 2 & 3 (Mikindani), 2007

YEAR	Trench	Size	Depth-Range	Pottery Total	Décorated	Shells	Bead	Bones	Lithic
2007	T1	1x4	0 – 50	88	2	3	0	0	0
	T2	3x3	0 – 50	92	32	8	0	0	2
	T3	2x4	0 – 60	77	2	0	0	0	0
	T4	4x1.5	0 – 60	89	6	7	0	0	3
	T5	3x3	0 – 50	183	5	9	0	0	1
	T6	3x1.5	0 – 60	62	1	1	0	0	0
	T7	2x2	0 – 50	54	4	0	0	0	0
	T8	2x3	0 – 50	193	23	2	0	0	0
	T9	2x3	0 – 40	373	26	4	4	0	0
	T10	1x3	0 – 80	559	76	11	1	0	1
	T11	1x2	0 – 40	93	14	0	0	0	1
	T12	1x2	0 – 50	87	7	1	0	0	0
TOTAL	**12**	**63.5m²**	**80 cm**	**1950**	**198**	**46**	**5**	**0**	**8**

1.6 Chikayowa (Mikindani), 2007

YEAR	Trench	Size	Depth-Range	Pottery Total	Décorrated	Shells	Bead	Bones	Lithic
2006	M6	2x3	0 - 70	1261	234	21	10	48	0
2007	T1	2x2	0 - 60	1985	110	0	6	127	0
	T2	1x2	0 - 60	455	53	1	2	5	0
TOTAL	**3**	**12m²**	**70cm**	**3701**	**397**	**22**	**18**	**180**	**0**

1.7 Nakatumbatu (Mikindani), 2007

YEAR	Trench	Size	Depth-Range	Pottery Total	Décor-rated	Shells	Bead	Bones	Lithic
2006	M7	1x1.6	0 - 50	658	2	0	0	0	0
2007	T1	1x2	0 - 60	170	1	1	0	0	2
TOTAL	**2**	**3.6m²**	**60CM**	**828**	**3**	**1**	**0**	**0**	**2**

1.8 Masakasa (Kilwa Island)

YEAR	Trench	Size	Depth-Range	Pottery Total	Décor-rated	Shells	Bead	Bones	Lithic
2007	T1	2x3	0 – 110	31	2	0	0	0	250
	T2	2x4	0 – 130	867	40	0	2	0	1448
	T3	3x1.5	0 – 110	919	33	0	0	0	94
	T3E	1x2	0 – 100	316	10	0	0	0	48
	T4	2x1.5	0 – 90	117	3	0	0	0	13
TOTAL	**5**	**23.5m²**	**130 cm**	**2250**	**88**	**0**	**2**	**0**	**1853**

1.9 Nguruni (Kilwa Island)

YEAR	Trench	Size	Depth-Range	Pottery Total	Décor rated	Shells	Bead	Bones	Lithic
2007	T1	3x3	0 - 80	437	3	0	0	0	0
	T2	2x2	0 - 100	56	3	0	0	0	1
	T3	1x2	0 - 100	41	0	0	0	0	0
	T4	1x2	0 - 110	187	5	0	1	0	4
	T5	1x2	0 - 80	413	6	0	0	0	0
TOTAL	**5**	**19m²**	**110 cm**	**1134**	**17**	**0**	**1**	**0**	**5**

Notes

T1, T2=Trench number; T = Total; D = Decorated pieces; T3E = trench 3 extension;

T1A = split trench into sector A, B… U6 = unit 6; TKR = Thickened rim;

TNR = Thinned rim; NTR = Neutral rim; LPI = Lip indented/indentations

CHD = Channeling decorations; PTD = Punctuates; STD = Stamps; SLA = iron slag

SHL = mollusks shells; BN = bones; BD = Beads; LIC = lithics

APPENDEX II

Frequencies of decorated pottery in different levels

Table 2.01   RUSHUNGI -KISWERE 2007 Pottery

Level	Depth	T 1		T 2		T 3		T 4	
	(cm)	T	D	T	D	T	D	T	D
1	0 - 10	30	3	23	0	16	0	150	21
2	10 - 20			107	6	9	3	287	45
3	20 - 30	85	9	46	2	12	1	337	49
4	30 - 40	19	1	2	0	0	0	45	8
5	40 - 50	4	0	0	0	0	0	23	7
6	50 - 60	2	0	0	0	0	0	12	2
7	60 - 70							8	2
Total		**140**	**13**	**178**	**8**	**37**	**4**	**862**	**134**

2.02a   MNANG'OLE - MCHINGA   (2007) –1: POTTERY

Level	Depth	T1		T2		T3		T3E		T4		T5		T5E	
	(cm)	T	D	T	D	T	D	T	D	T	D	T	D	T	D
1	0 - 10	194	18	420	2	15	0	11	0	10	0	8	4	25	1
2	10 - 20	607	124	71	29	23	7	53	9	184	22	385	49	7	0
3	20 - 30	211	27	16	5	112	19	107	5	157	8	1460	121	183	14
4	30 - 40	15	3	2	0	52	1	13	0	52	4	200	38	315	28
5	40 - 50	2	0	3	0	43	3	2	0	8	4	27	0	46	5
6	50 - 60									57	8	39	2	13	0
7	70 - 80									7	2	19	2	5	0
8	70 - 80											5	0	4	0
Total		**1029**	**172**	**512**	**36**	**245**	**30**	**186**	**14**	**475**	**48**	**2143**	**216**	**598**	**48**

184

2.02b MNANG'OLE - MCHINGA (2007) –2: POTTERY										
Level	Depth	T6		T7		T8		T9		T10	
	(cm)	T	D	T	D	T	D	T	D	T	D
1	0 - 10	3	0	11	0	14	1	5	0	35	0
2	10 - 20	0	0	73	16	38	3	14	2	31	0
3	20 - 30	106	9	63	10	98	7	216	10	4	0
4	30 - 40	83	10	1	1	20	16	44	0	2	0
5	40 - 50	11	1	0	0	2	0	13	0		
6	50 - 60							0	0		
7	70 - 80							0	0		
8	70 - 80										
Total		203	20	148	27	172	27	292	12	72	0

Notes

T1, T2…=Trench number; T = Total; D = Decorated pieces; T3E = trench 3 extension;

T1A = split trench into sector A, B… U6 = unit 6; TKR = Thickened rim;

TNR = Thinned rim; NTR = Neutral rim; LPI = Lip indented/indentations

CHD = Channeling decorations; PTD = Punctuates; STD = Stamps; SLA = iron slag

SHL = mollusks shells; BN = bones; BD = Beads; LIC = lit

2.03a MNAIDA 2 (2007) – 1: POTTERY															
Level	Depth	T1		T1A		T1B		T1C		T2A		T2B		T2C	
	(cm)	T	D	T	D	T	D	T	D	T	D	T	D	T	D
1	0 - 10	26	2	43	6	52	8	0	0	0	0	0	0	0	0
2	10 - 20	52	0	0	0	68	9	0	0	87	7	0	0	0	0
3	20 - 30	5	0					32	5	0	0	57	4	3	1
4	30 - 40	2	0											11	0
5	40 - 50	3	0												
6	50 - 60														
7	60 - 70														
8	70 - 80														
9	80 - 90														
Total		88	2	43	6	120	17	32	5	87	7	57	4	14	1

Level	Depth	T3A		T3B		T4		T4A		T4B		T5		T6		T 7	
	(cm)	T	D	T	D	T	D	T	D	T	D	T	D	T	D	T	D
1	0 - 10	0	0	0	0	13	0	0	0	0	0	60	2	14	0	23	1
2	10 - 20	0	0	0	0	33	3	0	0	0	0	82	3	14	0	30	3
3	20 - 30	5	1	0	0	24	2	0	0	0	0	36	3	37	1	1	0
4	30 - 40	1	0	18	0	13	1	9	0	1	1	36	0	7	0	0	0
5	40 - 50					3	0					4	0	4	0		
6	50 - 60					3	0					1	0				
7	60 - 70																
8	70 - 80																
9	80 - 90																
Total		6	1	18	0	89	6	9	0	1	1	219	8	76	1	54	4

Level	Depth	T8		T9		T10		T11	
	(cm)	T	D	T	D	T	D	T	D
1	0 - 10	130	10	81	3	41	3	18	0
2	10 - 20	175	13	278	14	50	6	30	0
3	20 - 30	58	3	140	13	22	5	22	5
4	30 - 40	10	0	40	3	2	0	15	2
5	40 - 50			11	3			15	2
6	50 - 60			4	2			2	0
7	60 - 70			3	0				
8	70 - 80			2	0				
9	80 - 90								
Total		373	26	559	38	115	14	102	9

2.05 KITERE (2007): POTTERY					
Level	**Depth**	**T1**		**T2**	
	(cm)	T	D	T	D
1	0 - 10	5	0	6	2
2	10 - 20	13	0	17	2
3	20 - 30	15	0	56	13
4	30 - 40	20	5	20	8
5	40 - 50	2	0	3	0
Total		**55**	**5**	**102**	**25**

2.06 CHIKAYOWA (MIKINDANI) 2007 & 2006 (M 6)							
Level	**Depth**	**T1**		**T2**		**T3 (M 6)**	
	(cm)	**T**	**D**	**T**	**D**	**T**	**D**
1	0 – 10	35	3	70	6	70	7
2	10 - 20	220	25	35	11	42	4
3	20 – 30	340	30	147	12	79	38
4	30 – 40	760	35	154	14	445	85
5	40 – 50	400	15	44	10	360	70
6	50 – 60	230	2	5	0	210	20
						55	10
Total		**1985**	**110**	**455**	**53**	**1261**	**234**

2.07 NAKATUMBATU (MIKINDANI) 2007 & 2006 (M 7)					
Level	**Depth**	**T1**		**T2 (M 7)**	
	(cm)	**T**	**D**	**T**	**D**
1	0 -10	11	1	0	0
2	10 - 20	3	0	14	0
3	20 - 30	89	0	640	2
4	30 - 40	66	0	4	
5	40 - 50	1	0		
Total		**170**	**1**	**658**	**2**

Notes

T1, T2…=Trench number; T = Total; D = Decorated pieces; T3E = trench 3 extension;

T1A = split trench into sector A, B… U6 = unit 6; TKR = Thickened rim;

TNR = Thinned rim; NTR = Neutral rim; LPI = Lip indented/indentations

CHD = Channeling decorations; PTD = Punctuates; STD = Stamps; SLA = iron slag

SHL = mollusks shells; BN = bones; BD = Beads; LIC = lithics

2.08a MIKINDANI (2006) – 1: POTTERY

Level	Depth	M1A		M1B		M1C		M2A		M2B		M3		M4	
	(cm)	T	D	T	D	T	D	T	D	T	D	T	D	T	D
1	0 - 10	5	1	4	0	11	0	46	2	0	0	75	15	44	8
2	10 - 20	7	1	4	0	2	0	92	11	18	6	78	24	32	4
3	20 - 30	2	0	2	0	1	0	15	1	8	5	102	31	71	3
4	30 - 40							3	0	0	0	45	2	81	4
5	40 - 50											5	0	7	8
6	50 - 60													5	0
7	60 - 70													4	2
8	70 - 80													1	1
Total		**14**	**2**	**10**	**0**	**14**	**0**	**156**	**14**	**26**	**11**	**305**	**72**	**245**	**30**

2.08b MIKINDANI (2006) – 2: POTTERY

Level	Depth	M5		M6 (U6)		M7 (U7)	
	(cm)	T	D	T	D	T	D
1	0 – 10	0	0	70	7	0	0
2	10 - 20	4	0	42	4	14	2
3	20 – 30	27	8	79	38	640	0
4	30 – 40	37	14	445	85	4	0
5	40 – 50	10	1	360	70	0	0
6	50 – 60	1	0	210	20	0	0
7	60 – 70	0	0	55	10	0	0
8	70 – 80	0	0				
Total		**79**	**23**	**1261**	**234**	**658**	**2**

2.09 Kilwa Island - MASAKASA (2007) POTTERY

Level	Depth	TI		T2		T3		T3E		T4	
	(cm)	T	D	T	D	T	D	T	D	T	D
1	0 - 10	1	1	70	2	10	0	4	1	36	0
2	10 - 20	2	1	265	13	2	0	160	4	45	3
3	30 - 40	23	0	160	15	155	6	104	2	23	0
4	40 - 50	5	0	89	3	613	24	104	2	13	0
5	50 - 60			240	5	40	2	16	0	3	0
6	60 - 70			37	0	50	1	30	3	2	0
7	70 - 80			6	2	36	0	2	0	0	0
8	80 - 90					3	0			2	0
9	90 - 100					7	0				
10	100 -110					3	0				
Total		31	2	867	40	919	33	420	12	124	3

2.10 Kilwa Island - NGURUNI (2007) POTTERY

Level	Depth	T1		T2		T3		T4		T5	
	(cm)	T	D	T	D	T	D	T	D	T	D
1	0 - 10	2	0	25	2	4	0	6	0	232	2
2	10 - 20	120	1	25	1	4	0	150	3	3	0
3	20 - 30	186	0	4	0	9	0	31	2	133	4
4	30 - 40	87	2	2	0	19	0			33	0
5	40 - 50	18	0			5	0			12	0
6	50 - 60	24	0								
Total		437	3	56	3	41	0	187	5	413	6

APPENDEX III

Variations in Pottery rim-forms, designs and frequency of other artefacts in different levels and trenches.

Table 3.1 POTTERY DESIGNS AND OTHERS– MASAKASA												
Trench	Depth	TKR	TNR	NTR	LPI	CHD	PTD	STD	SLA	SHL	BN	BD	LIC
1	0 - 10												
	10- 20												
	20 - 30			2									4
	30 - 40												13
	40 - 50												
	50 - 60												153
	60 - 70												80
	Total			**2**									**250**
2	0 - 10	0	0	1	0	2	0	0	1				3
	10- 20	0	8	16	0	4	3	0	3				5
	20 - 30	0	10	5	0	11	0	1	11				7
	30 - 40	1	3	3	0	1	0	0	12				20
	40 - 50	3	4	4	2	2	1	2	23			2	33
	50 - 60	0	0	1	0	0	0		10				11
	60 - 70	1	0	1	2	3	0						103
	70 - 80			0	0								330
	80 - 90												440
	90 - 100												400
	100 - 110												103
	110 - 120												29
	120 - 130												14
	Total	**5**	**25**	**31**	**4**	**23**	**4**	**3**	**60**			**2**	**1498**
3	0 - 10												
	10 20												
	20 - 30							1					
	30 - 40	0	4	2	0	0	0	8					
	40 - 50	0	7	15	1	10	4	6					
	50 - 60	0	4	2	1	2	0	0					2
	60 - 70	7	1	6	2	2	2	0					2
	70 - 80	2	0	0	0	0	0	0					6
	80 - 90												90
	90 - 100												36
	100 -110												6
	Total	**6**	**16**	**25**	**4**	**14**	**6**	**15**					**142**

4	30 - 40	0	2	0									
	40 - 50	0	0	2				1					
	50 - 60	1	0	0									
	60 - 70												
	70 - 80											2	
	80 - 90											11	
	Total	**1**	**2**	**2**				**1**				**13**	

<table>
<tr><td colspan="14">3.2 POTTERY DESIGNS & OTHERS- NGURUNI</td></tr>
</table>

Trench	Depth (cm)	TKR	TNR	NTR	LPI	CHD	PTD	STD	SLA	SHL	BN	BD	LIC
1	0 - 10												
	10 - 20												
	20 - 30	2	7	2	0	0	0	0	4				
	30 - 40	2	0	3	0	2	0	0	1				
	40 - 50	0	0	2									
	Total	**4**	**7**	**7**	**0**	**2**	**0**	**0**	**5**				
2	0 - 10	0	1	0	0	2	0	0	0	0	0	0	0
	10 -20		0			1	0		1				1
	20 - 30												
	30 - 40	0	0	0	0	0	0	0	0	0	0	0	0
	Total		**1**			**3**			**1**				**1**
3	0 - 10												
	10 - 20												
	20 - 30		1										
	30 - 40		0										
	40 - 50		1										
	Total		**2**										
4	0 - 20	0	0	0	0	0	0		0	0	0	0	0
	20 - 30	1	0	2	0	1	2		2	0	0	1	4
	30 - 40	0	0	0	0	2	0		0	0	0	0	0
	Total	**1**	**0**	**2**	**0**	**3**	**2**		**2**	**1**	**0**	**1**	**4**
5	0 - 20	0	7	7	0	2							
	20 - 30	0	0	0	0	0							
	30 - 40	0	1	5	0	1	2						
	40 - 50	0	2	1	0	0				0			
	50 - 60									1			
	Total	**0**	**10**	**13**	**0**	**3**	**2**			**1**			

<u>Notes</u>

T1, T2…=Trench number; T = Total; D = Decorated pieces; T3E = trench 3 extension;

T1A = split trench into sector A, B… U6 = unit 6; TKR = Thickened rim;

TNR = Thinned rim; NTR = Neutral rim; LPI = Lip indented/indentations

CHD = Channeling decorations; PTD = Punctuates; STD = Stamps; SLA = iron slag

SHL = mollusks shells; BN = bones; BD = Beads; LIC = lithics

3.3a Pottery designs and Others - MNANG'OLE (2007):

Trench	Depth (cm)	TKR	TNR	NTR	LPI	CHD	PTD	STD	SLA	SHL	BN	BD	LIC
1	0 - 10	0	4	2	3	2	8	5	0	152			
	10 - 20	0	32	19	11	15	42	55	25	35			
	20 - 30	1	9	3	3	9	5	10	13	10			
	30 - 40	0	0	0	0	0	1	1	0	0			
	Total	**1**	**45**	**24**	**17**	**26**	**56**	**71**	**38**	**197**			
2	0 - 10	0	0	0	0	0	0	2					
	10 - 20	0	9	0	6	11	3	8					
	20 - 30	1	1	1	3	0	1	1					
	Total	**1**	**10**	**1**	**9**	**11**	**4**	**11**					
3	0 - 10	0	0	4	0	0	0	0	0	25	0	0	0
	10 - 20	1	4	2	2	0	1	4	0	0	0	0	0
	20 - 30	2	11	1	9	5	2	4	7	0	0	1	0
	30 - 40	0	0	1	0	0	0	1	1	1	0	0	1
	40 - 50	0	4	0	0	2	1	0	0	0	0	0	0
	Total	**3**	**19**	**8**	**11**	**7**	**4**	**9**	**8**	**26**	**0**	**1**	**1**
3 E	0 - 10	0	1	0	0	0	0	0	0	0	0	0	0
	10 - 20	0	2	1	2	0	4	3	0	0	0	0	0
	20 - 30	0	4	3	1	0	0	4	0	1	0	0	0
	30 - 40	0	0	0	0	0	0	0	0	0	0	0	1
	40 - 50	0	0	0	0	0	0	0	0	0	0	0	2
	Total	**0**	**7**	**4**	**3**	**0**	**4**	**7**	**0**	**1**	**0**	**0**	**3**
4	0 - 10	0	0	0	0	0	0	0	0			0	0
	10 - 20	0	2	1	2	6	5	11	0			1	0
	20 - 30	3	4	2	2	2	2	4	9			0	0
	30 - 40	1	2	1	0	1	1	1	3			0	0
	40 - 50	0	3	0	1	0	1	3	0			0	1
	50 - 60	0	2	0	1	1	2	5	0			0	5
	60 - 70	0	0	1	0	0	1	0	0			0	0
	Total	**4**	**13**	**5**	**6**	**10**	**12**	**24**	**12**			**1**	**6**
5	0 - 10	0	1	0	1	1	1	2	5		0		0
	10 - 20	0	6	4	2	2	5	42	0		0		1

Trench	Depth	TKR	TNR	NTR	LPI	CHD	PTD	STD	SLA	SHL	BN	BD	LIC
	20 - 30	0	0	0	35	3	35	87	0		15		4
	30 - 40	1	6	4	7	3	7	28	0		0		0
	40 - 50	1	0	0	0	0	0	0	8		0		1
	50 - 60	2	0	0	0	0	2	0	11		0		2
	60 - 70	1	0	0	0	1	1	0	2		0		0
	70 - 80	0	0	0	0	0	0	0	2		0		4
	Total	**5**	**13**	**8**	**45**	**10**	**51**	**159**	**28**		**15**		**12**

3.3b   Mnangole – Mchinga 2007 (Continued)

Trench	Depth	TKR	TNR	NTR	LPI	CHD	PTD	STD	SLA	SHL	BN	BD	LIC
5E	0 - 10	0	4	0	0	0	0	1	1		0		0
	10 - 20	0	0	0	0	0	0	0	0		0		0
	20 - 30	0	4	6	2	0	3	11	0		0		0
	30 - 40	0	11	4	6	6	6	14	5		5		0
	40 - 50	3	0	0	0	0	5	0	8		0		1
	50 - 60	0	0	0	0	0	0	0	1		0		5
	60 - 70	2	0	0	0	0	0	0	4		0		6
	70 - 80	0	0	0	0	0	0	0	0		0		1
	Total	**5**	**19**	**10**	**8**	**6**	**14**	**26**	**19**		**5**		**13**
6	0 - 10												
	10 - 20												
	20 - 30		3	2	2	1	3	5	0	0	0	1	0
	30 - 40		1	2	1	2	2	6	0			0	
	40 - 50		0	0	0	0	0	1	4		0	0	
	Total		**4**	**4**	**3**	**3**	**5**	**12**	**4**			**1**	
7	0 - 10	0	0	1	0	0	0	0		0			
	10 - 20	0	5	1	0	0	1	15		5			
	20 - 30	0	5	1	1	4	3	3		0			
	30 - 40	1	0	0	0	1	0	0		0			
	Total	**1**	**10**	**3**	**1**	**5**	**4**	**18**		**5**			
8	0 - 10	0	1	1	1	0	1	0	0	0	0	0	0
	10 - 20	0	0	0	0	0	0	3	0				0
	20 - 30	1	3	2	1	5	1	1	5				0
	30 - 40	7	0	0	0	2	1	13	2	0	0	0	1
	Total	**8**	**4**	**3**	**2**	**7**	**3**	**17**	**7**				**1**
9	0 - 10	0	0		0	0	0	0	0				
	10 - 20	0	0		0	0	0	2	1				
	20 - 30	1	3		1	3	2	5	41				
	30 - 40	3	0		0	0	0	0	123				
	40 - 50	0	0		0	0	0	0	12				2
	50 - 60								3				0
	60 - 70												2

	70 - 80												0
	Total	**4**	**3**		**1**	**3**	**2**	**7**	**180**				
10	0 - 10									42	1		
	10 - 20									82	1		
	20 - 30									22	3		
	30 - 40									7	0		
	Total									**153**	**5**		

<u>Notes</u>

T1, T2…=Trench number; T = Total; D = Decorated pieces; T3E = trench 3 extension; LIC = lithics

T1A = split trench into sector A, B… U6 = unit 6; TKR = Thickened rim; TNR = Thinned rim

; NTR = Neutral rim; LPI = Lip indented/indentations CHD = Channeling decorations; BD = Beads;

PTD = Punctuates; STD = Stamps; SLA = iron slag SHL = mollusks shells; BN = bones;

3.4 POTTERY DESIGNS - MNAIDA 2 & 3 (2007)

Trench	Depth (cm)	TKR	TNR	NRT	LPI	CHD	PTD	STD	SLA	SHL	BN	BD	LIC
1	0 - 10			0			2			0			
	10 - 20			5			0			0			
	20 - 30			0			0			0			
	30 - 40			0			0			3			
	40 - 50									+			
	Total			**5**			**2**			**3**			
2	0 - 10 (1)	0	0	0	0	0	0	0	0	1	0	0	0
	0 - 10(2)		0	0	0	0	0			0			0
	10 - 20(1)		1	2	0	1	0			0			0
	10 - 20(2)		0	0	0	2	0			0			1
	20 - 30(1)		0	2	0	1	4			5			0
	20 - 30(2)		0	1	0	1	0			+			1
	30 - 40(1)		0	0	0	0	0			+			0
	30 - 40(2)	0	1	0	0	1	0	0	0	+	0	0	0
	Total		**2**	**5**	**0**	**6**	**4**			**6**			**2**
3	0 - 10		0	1		1							
	10 - 20		0	0		0							
	20 - 30		0	1		0							
	30 - 40		1	1		0							
	40 - 50		0	1		0							

Trench	Depth (cm)	TKR	TNR	NTR	LPI	CHD	PTD	STD	SLA	SHL	BN	BD	LIC
	50 - 60												
	Total		**1**	**4**		**1**							
4	0 - 10		0	0		0	0						0
	10 - 20		1	0		0	3						0
	20 - 30		1	1		1	1						0
	30 - 40		0	0		1	0	0	0	0	0	0	1
	40 - 50		0	0		0	0					0	2
	50 - 60												
	Total		**2**	**1**		**2**	**4**						**3**
5	0 - 10	0	0	3	0	1	1	0	0	0	0		
	10 - 20	0	0	8	1	0	2	0	0	9	0		
	20 - 30			0	0	0	0						
	30 - 40			0	0		0						0
	40 - 50			0	0		0					0	1
	Total	**0**	**0**	**11**	**1**	**1**	**3**	**0**	**0**	**9**			**1**
6	0 - 10												
	10 - 20		0	1	0	0	0	0	0	0			
	20 - 30		1	0	0	1	0	0	0	0			
	30 - 40		0	0	0	0	0	0	0	0			
	40 - 50		0	0	0	0	0	0	0	0			
	50 - 60		0	0	0	0	0	0	1	1			
	Total		**1**	**1**	**0**	**1**	**0**	**0**	**1**	**1**			

Mnaida 2 & 3, 2007 (continued)

Trench	Depth (cm)	TKR	TNR	NTR	LPI	CHD	PTD	STD	SLA	SHL	BN	BD	LIC
7	0 - 10												
	10 - 20	0	0	2	0	0	0	1					
	20 - 30	0	2	3	0	0	0	3					
	30 - 40												
	40 - 50												
	Total		**2**	**5**				**4**					
8	0 - 10		0	3		0	0	6	0	2	0		
	10 - 20(1)		1	1		0	7	1		0	0		
	10 - 20(2)		1	0		0	0	2		0	0		
	20 - 30(2)		2	0		0	3	2		0			
	20 - 30(3)		0	0		1	0	0		0			
	30 - 40		0	0		0	0	0		0			
	40 - 50		0	0		1	0	0		0			
	Total		**4**	**4**		**2**	**10**	**11**		**2**	**0**		
9	0 - 10		0	0		3	7	0	0	0	0	0	

Trench	Depth	TKR	TNR	NTR	LPI	CHD	PTD	STD	SLA	SHL	BN	BD	LIC
	10 - 20		0	4		2	10	1	0	4	0	1	
	20 - 30		0	1		0	1	1				3	
	30 - 40		3	0		0	0	0					
	Total		**3**	**5**		**5**	**18**	**2**	**0**	**4**	**0**	**4**	
10	0 - 10		3	1		0	0	0	4	0		1	0
	10 - 20		0	0		3	4	7	0	6		0	1
	20 - 30		2	3		1	5	6	0	2		0	0
	30 - 40		0	2		1	2	0	1	3		0	0
	40 - 50		0	0		0	3	0	0	0		0	0
	50 - 60		0	0		0	2	0	0	0		0	0
	Total		**5**	**6**		**5**	**16**	**13**	**5**	**11**		**1**	**1**
11	0 - 10			3		1	2	0					0
	10 - 20			1		0	3	2					0
	20 - 30			1		0	3	2					1
	30 - 40												
	Total			**5**		**1**	**8**	**4**					**1**
12	0 - 10								1				
	10 - 20								0				
	20 - 30	2	3	0	1	1	3		0				
	30 - 40	0	3	0	0	2	2		0				
	Total	**2**	**6**	**0**	**1**	**3**	**5**		**1**				

3.5 POTTERY DESIGNS & OTHERS – KITERE (2007)													
Trench	Depth	TKR	TNR	NTR	LPI	CHD	PTD	STD	SLA	SHL	BN	BD	LIC
	(cm)												
1	0 - 10												
	10 - 20												
	20 - 30												
	30 - 40	0	3	0	0	2	1	2					
	40 - 50	0	0	0	0	0	0	0	0	0	0	0	14
	50 - 60											0	8
	Total	**0**	**3**	**0**	**0**	**2**	**1**	**2**					**22**
2	0 - 10												0
	10 - 20												2
	20 - 30												6
	30 - 40												37
	40 - 50	1	0	3	0	1	7						17
	50 - 60												10
	Total	**1**	**0**	**3**	**0**	**1**	**7**						**72**

3.6 POTTERY DESIGNS & OTHERS- CHIKAYOWA 2007

Trench	Depth (cm)	TKR	TNR	NTR	LPI	CHD	PTD	STD	SLA	SHL	BN	BD	LIC
1	0 - 10	0	0	4	0	1	0	2			0	0	
	10 - 20	0	7	13	5	5	3	14			6	1	
	20 - 30	0	12	19	8	3	2	11			15	3	
	30 - 40	0	18	22	6	9	3	20			60	2	
	40 - 50	0	10	10	3	1	4	7			26	0	
	50 - 60	0	1	4	1	0	0	1			20	0	
	Total	**0**	**48**	**72**	**23**	**19**	**12**	**55**			**127**	**6**	
2	0 - 10	0	3	0	0	0	2	4		1	1	1	
	10 - 20	0	8	0	0	0	7	4		0	0	0	
	20 - 30	0	9	4	1	0	3	9		0	3	0	
	30 - 40	0	3	4	1	0	0	0		0	1	1	
	40 - 50	0	5	4	0	4	6	0		0	0	0	
	Total	**0**	**28**	**12**	**2**	**4**	**18**	**17**		**1**	**5**	**2**	

3.7 POTTERY DESIGNS & OTHERS- NAKATUMBATU 2007

Trench	Depth (cm)	TKR	TNR	NTR	LPI	CHD	PTD	STD	SLA	SHL	BN	BD	LIC
1	0 - 10	0	1			1							0
	10 - 20	0	0			0							0
	20 - 30	0	3			0							0
	30 - 40	0	5			0							0
	40 - 50	0	0			0							2
	Total	**0**	**9**			**1**							**2**

3.8 POTTERY DESIGNS AND OTHERS - RUSHUNGI

Trench	Depth (cm)	TKR	TNR	NTR	LPI	CHD	PTD	STD	SLA	SHL	BN	BD	LIC
1	0 - 10												
	10 - 20	0	0	0	2	1	1	1		0	0	0	0
	20 - 30	0	0	0	3	5	0	3		370	0	2	0
	30 - 40				0	1	0	0		740	0	0	3
	40 - 50					0	0	0		80	0	0	1
	50 - 60					0	0	0		0	0	0	0
	Total					**7**	**1**	**4**		**1190**		**2**	**4**

Trench	Depth	TKR	TNR	NTR	LPI	CHD	PTD	STD	SLA	SHL	BN	BD	LIC
2	0 - 10	0	0	0	0	0	0	0	0	70	0	0	1
	10 - 20					2	0	4	0	136			4
	20 - 30					1	0	1	0	78			4
	30 - 40					0	0	0	0	0			0
	Total					**2**	**0**	**5**	**0**	**284**			**9**
3	0 - 10					0	0	0					
	10 - 20					1	0	2					
	20 - 30					0	0	1					
	Total					**2**	**0**	**2**					
4	0 - 10	0	0	0	9	18	0	3	0	0	0	1	
	10 - 20				14	26	8	4				5	
	20 - 30				17	25	0	0				32	
	30 - 40				2	8	0	0				3	
	40 - 50				**1**	**4**	**0**	**3**				**8**	
	50 - 60				**0**	**1**	**0**	**0**					
	60 - 70				**0**	**1**	**0**	**1**					
	Total					**77**	**8**	**5**				**49**	

3.9 POTTERY DESIGNS - MIKINDANI (2006)

Trench	Depth	TKR	TNR	NTR	LPI	CHD	PTD	STD	SLA	SHL	BN	BD	LIC
	(cm)												
1	0 - 10(A)					1							
(M1)	0 - 10 (B)									7			
	0 - 10(C)	0	1	0	0	0	0	0	0	2	0	0	0
	10 - 20(A)							1		3			
	10 - 20(B)												
	10 - 20(C)									3			
	20 - 30(A)												
	20 - 30(B)												
	20 - 30(C)	0	0	1	0	0	0	0	0	2	0	0	0
	Total		**1**	**1**		**1**		**1**		**17**			

Mikindani 2006 (continued):

Mikindani 2006 (continued):

Trench	Depth	TKR	TNR	NTR	LPI	CHD	PTD	STD	SLA	SHL	BN	BD	LIC
2	0 - 10(A)	0	4	1	0	1	0	1	0	27			
(M2)	0 - 10(B)	0	0	0	0	0	0	0	0	0			
	10 - 20(A)	0	0	2	0	5	4	2	0	48			
	10 - 20 (B)	0	0	2	0	0	0	5	0	8			
	20 - 30(A)	0	1	1	0	1	0	0	0	5			
	20 - 30(B)	0	0	0	0	1	0	4	0	2			
	Total		**5**	**6**	**0**	**8**	**4**	**12**	**0**	**90**			
3	0 - 10	0	1	10	0	1	12	3					
(M3)	10 - 20	0	0	0	2	4	7	8					
	20 - 30	0	0	0	0	0	0	0					
	30 - 40	2	2	5	0	0	0	0	0	0	0	2	
	Total	**2**	**3**	**15**	**2**	**5**	**19**	**9**				**2**	
4	0 - 10	0	0	0	0	3	0	5	0	15	0		
(M4)	10 - 20	0	0	0	0	4				5			
	20 - 30	1	0	4	0	0							
	30 - 40	0	0	3	0	3							+
	40 - 50	0	0	0	0	8							+
	50 - 60					0							+
	60 - 70					2							+
	70 - 80					1							
	Total	**1**	**0**	**7**	**0**	**21**		**5**		**20**			
5	10 - 20												
(M5)	20 - 30				0	8	0	0					
	30 - 40				0	11	0	3					+
	40 - 60				0	1	0	0					+
	Total				**0**	**20**	**0**	**3**					
6	0 - 10						2	5					
(M6)	10 - 20						1	3				1	
	20 - 30				4	6	6	23				5	
	30 - 40				11	16	0	12	0	0	3	2	
	40 - 50				15	9	14	7	0	8	15	2	
	50 - 60	0	7	0	5	8	2	4	0	7	21	0	
	60 - 70				2					3	9	0	
	Total		**7**		**37**	**39**	**25**	**54**		**18**	**48**	**12**	

Notes

T1, T2…=Trench number; T = Total; D = Decorated pieces; T3E = trench 3 extension; T1A = split trench into sector A, B… U6 = unit 6; TKR = Thickened rim;

TNR = Thinned rim; NTR = Neutral rim; LPI = Lip indented/indentations;

CHD = Channeling decorations; PTD = Punctuates; STD = Stamps; SLA = iron slag;